Substrate Surface Preparation Handbook

For a complete listing of titles in the
Artech House Applied Photonics Series,
turn to the back of this book.

Substrate Surface Preparation Handbook

Max Robertson

ARTECH HOUSE

BOSTON | LONDON
artechhouse.com

Library of Congress Cataloging-in-Publication Data
A catalog record for this book is available from the U.S. Library of Congress.
British Library Cataloguing in Publication Data
A catalogue record for this book is available from the British Library.

Cover design by Vicki Kane

ISBN 13: 978-1-60807-213-2

© 2012 ARTECH HOUSE
685 Canton Street
Norwood, MA 02062

The following figures are from photographs published with kind permission of Logitech Ltd, Old Kilpatrick, Glasgow: 1.1, 1.2, 2.4, 2.7–2.9, 3.1, 3.4, 4.1, 4.2, 4.4, 4.5, 4.7–4.10, 5.1, 5.3, 6.1, 6.3, 6.4, 7.1, 7.2, 9.1, 11.1, 11.2.

All rights reserved. Printed and bound in the United States of America. No part of this book may be reproduced or utilized in any form or by any means, electronic or mechanical, including photocopying, recording, or by any information storage and retrieval system, without permission in writing from the publisher.
All terms mentioned in this book that are known to be trademarks or service marks have been appropriately capitalized. Artech House cannot attest to the accuracy of this information. Use of a term in this book should not be regarded as affecting the validity of any trademark or service mark.

10 9 8 7 6 5 4 3 2 1

Contents

	Foreword	**11**
	Preface	**13**
1	**Introduction**	**15**
1.1	Choosing a Process	15
1.2	Definitions of Processes Used in This Book	17
1.3	Lapping, Grinding, and Polishing Abrasives	19
2	**Preparation: Before the Start**	**21**
2.1	Plates and Measurement	21
2.1.1	Plate Measurement	22
2.1.2	Maintaining Plate Shape	23
2.2	Lapping Plates	23
2.2.1	Glazing	25

2.3	Polishing Plates	26
2.4	Polishing Surfaces: Care and Conditioning	27
2.5	Baseplates for Polishing	29
2.5.1	Baseplate Materials	29
2.6	The Use of Smoothing Blocks	30
2.7	Jigs	33
2.7.1	When to Use a Jig	33
2.7.2	Jig Balance	34
2.7.3	Jig Maintenance	34
2.8	Sample Mounting	36
2.8.1	Vacuum Mounting	36
2.8.2	Wax Mounting	37
2.8.3	Automated Bonding	39
2.8.4	Evaporated Wax Films	40
2.8.5	Surface Tension Mounting	41
2.8.6	Epoxy Bonding	42
2.9	Sample Viewing and Assessment	42
2.10	Plate and Sample Flatness Control	44
2.10.1	Wafer Distortion	45
2.11	Conclusion	46
	References	47
3	**Lapping**	**49**
3.1	The Lapping Process	49
3.1.1	If the Stock Removal Is Too Slow	50
3.1.2	If the Stock Removal Is Too Fast	50
3.2	Plate Shape Monitoring	51
3.3	Scratching	52

3.4	Smoothing	54
	References	56

4	**Polishing**	**57**
4.1	Introduction	57
4.2	Sample Load	59
4.3	Abrasives	61
4.4	Edge Polishing	62
4.5	Slurry Flow Rate	62
4.6	Grit Sizes	62
4.7	Aligning the Sample	62
4.8	The Polishing Run	66
4.8.1	Before the Start	66
4.8.2	Monitoring Progress	66
4.9	Jig Rotation	67
4.10	Sample Surface Shape and In-Process Alignment	67
4.11	Chemical Polishing	69
4.12	Chemomechanical Polishing	73
4.13	Fluid Jet Polishing: Future Developments	75
	References	75

5	**Specific Processes and Materials**	**77**
5.1	Geology	77
5.2	Hard Materials	81

| 5.2.1 | Lapping Hard Materials | 82 |
| 5.2.2 | Polishing Hard Materials | 83 |

| 5.3 | Water-Soluble Materials | 86 |

5.4	Electro-Optic Materials	88
5.4.1	Infrared and Electro-Optic Materials	88
5.4.2	Processing Infared and Electro-Optic Materials	90
	References	91

6 Specialized Techniques 93

6.1	Diamond Machining: Introduction	93
6.1.1	Diamond Machining of Ductile Materials	94
6.1.2	Diamond Machining of Brittle Materials	95

6.2	Sawing	96
6.2.1	Wire Saws	97
6.2.2	High-Speed Saws	98
6.2.3	Annular Saws	100
	References	102

7 Surface Finish 103

| 7.1 | The Lapped Surface Finish | 103 |

| 7.2 | Subsurface Damage | 104 |

7.3	Understanding Surface Finish	105
7.3.1	Cutoff	106
7.3.2	Stylus Radius	106
	Reference	110

8 Optics 111

| 8.1 | Glass | 112 |

| 8.2 | Processing with Pitch | 113 |

8.3	Pitch Alternatives	117
8.4	Spherical Surfaces	118
8.5	Blocking Spherical Components	119
8.6	Specifying Diamond Tooling	121
8.7	Testing of Optical Components	122
	References	123

9 Semiconductor Device Deconstruction — 125
| | References | 132 |

10 Metallurgical Polishing and Microscopy — 133
10.1	Processing	133
10.1.1	Process Stages	134
10.2	Examination	134
10.3	Microscope Setup	135
	References	136

11 Laboratory Setup — 137
11.1	Equipment Locations	137
11.2	Laboratory Layout and Dimensions	138
11.3	Optimizing the Process Route	141
11.4	Sample Cleaning	142
11.5	Safety Regulations	144
11.6	Lab Environment	145
11.7	Consumables	145
	References	146

12	**Using Interferometry**	**147**
12.1	Basic Principles	147
12.2	Analysis of Fringe Patterns	149
12.3	Normal and Grazing Incidence	151
12.4	Introducing the Workshop Interferometer	153
	References	155

Bibliography — **157**

Glossary — **159**
References — 178

Appendix: The Workshop Grazing Incidence Interferometer — **181**

A.1	Optical Path	181
A.2	Stage Mirrors and Adjustment	184
A.3	Hardware and Detailed Drawings	188

About the Author — **189**

Index — **191**

Foreword

Polishing has come of age. Starting from the loin-clothed individual cleaning off the excess glass on the base of a blown-glass work of art using a large rotating stone, it has progressed to being now used in large-scale processes. Processes with atomic-scale precision, over many cycles, on the large-diameter silicon wafers are used in the mass production of electronic integrated circuits. In such areas of application, few nanometer-thick layers of precious metals on electronic devices must be polished to atomic tolerances, so that the next structured layer of the device can be registered and deposited to the required accuracy.

However, the basic principles of using a rotating wheel, along with random motion of the sample, still serve both ends of this spectrum of sophistication. More than ever, polishing has become a production tool, and as such, it influences the direction taken by researchers everywhere. The ability to planarize what has become a huge stack of increasingly fine layers in a semiconductor device, with ever-increasing accuracy, enables devices of progressively greater increased complexity and sophistication to be dreamt of, conceived, and then realized. In addition, a wider and wider range of materials can be processed—right up to the hardest of them all—diamond.

When Bob Wilson, a researcher at the University of Glasgow, wanted to thin and then finish slices of cadmium sulphide to have a perfectly polished surface so as to produce a prototype device in 1965, he found a distinct lack of available equipment and technical advice. So, he founded a company in his garage in Alexandria (the Scottish one) to address this—and started a voyage which progressed through the many possible applications, ranging from semiconductors to geology and on to optics and electro-optics. The company developed and grew, and while having several owners, still maintained the straightforward principles of technology transfer and versatility over a worldwide market, in a full range of processing equipment. Max Robertson spent over a decade as development manager of this company and, although now retired, viewed the changing face of the technology very much from the point of view of the people involved.

In charge of the training department, he increasingly saw that research assistants as trainees were giving way to technicians, who, by definition, came from a wide range of backgrounds with little or no prior knowledge of polishing technology. Aimed at providing a "quick start" route to successful polishing, this book develops what has sometimes been described as a black art, into a logical process, and continues by unveiling the secrets of device delamination and working with ultrathin layers.

Richard M. De La Rue
Professor of Opto-Electronics
University of Glasgow
Glasgow, Scotland
November 2011

Preface

During 20 enjoyable and busy years at Logitech Ltd., I have become increasingly impressed by the simplicity and robustness of the technology. At the same time, because of the large number of pitfalls which accompanied its use, there seemed to be a real need to supplement the company's in-house training courses with an easy to follow text. This would enable trainees to arrive ready to start on their own very important samples, and those who supervise them, to understand why sometimes the process takes longer than expected. Potential employees in the sample preparation, optics, and semiconductor industries could arrive primed and prepared to discuss an unfamiliar area. The original company concept of providing a process route and technology for new materials finishing, is as valid today as it was in 1965, and I am very grateful for Logitech's support in this project.

The reader will find that each chapter, and in fact the book, is laid out to give the simple technology aimed at problem solving at the start, followed later by more academic points. This provides the researcher, supervisor, and technician with a concise introduction to the subject and facilitates use of the index. In addition, any really relevant detail topics (mainly those areas I personally found confusing and never had time to investigate) are explained in the Glossary. Any academic references are at the end of the relevant chapter.

Because of the similarities in technology, it is very possible that employees from research laboratories may move onto optical and semiconductor preparation industries. The section on optics processing has been extended, thanks to the contribution from Gary Thomas, to accommodate this and give a flavor of the pressures generated by a production environment. Perhaps some interested readers may also be persuaded to take up the challenges posed by electro-optic and semiconductor materials.

Throughout the text, examples of the usefulness of the grazing incidence interferometer are given, and designs for a very simple, cost-effective instrument are included in the Appendix.

Many thanks are due to Richard De La Rue for his kind foreword, to Logitech Ltd., and to all my colleagues there for putting up with me for 20 years, and to my wife, Winifred, for putting up with me for twice as long! Especially, my thanks go to Jim McAneny for his patience in checking illustrations and text and for the many days he spent training the author on his first arrival at the company. I am also very grateful to Gary Thomas and John Winfield, for their help.

1

Introduction

1.1 Choosing a Process

Very often there is no choice. For students, technicians, or supervisors who are handed a sample of an unknown material and asked to produce an infinitely thin slice, with an impossibly good finish, in an unreasonably short time, using the ancient kit available; choosing a process route can be a bit academic.

On the other hand, researchers who have just developed, or are hoping to develop, a new material for a specific application which demands an atomic-level finish as a *substrate* for subsequent *epitaxial* growth, then a high level of planning and investment may be required. These extreme situations hint at the scope of contents, and the prospective audience for this book as explained in the Foreword and Preface. The words in italics above, and from now on, are elaborated in the Glossary.

No two research or test samples are ever the same. Certain hardware and materials are essential, and no amount of ingenuity will get over a scratching problem, if the polishing abrasive used came out of the scourer carton under the sink. This type of abrasive has one purpose only: try it and you will be convinced instantly that the three essentials for the job in hand are: DISCIPLINE, PATIENCE, and GOOD EQUIPMENT. This sounds incredibly dull, but shortcuts can only be taken if the practitioner has enough experience to fully understand the consequences. The longest chapter in this book is on PREPARATION. Anticipation is another keyword which springs to mind.

It is assumed throughout this book, that certain pieces of lab equipment are available, even if they may have to be borrowed, including microscope, stylus profiling and surface-finish measuring instruments, and interferometers. This book emphasizes the practice of using interferometry as a tool for sample assessment and alignment, simply because it has proved its worth over many years in laboratory and workshop environments. In particular, the grazing incidence interferometer was the only instrument in its development year in the company for whom the author worked, that had a regular stream of visitors from workshop to laboratory demanding its use. Faced with a new sample, the process to be chosen by the reader will rapidly become apparent with progress through this book.

Representing wavelength, in this case of visible light, lambda (Figure 1.1) signifies that the processes of lapping and polishing are fundamental to the preparation of the best possible surface finishes. Measurements of these surfaces often use light waves. The advent of cost-effective *laser diode modules* has made it possible to create economical new instruments which can manipulate light in such a way as to facilitate sample alignment and analysis at all stages of processing. These tests, which used to take up so much of the technician's time and cause so much frustration, can now be much simpler, making it possible to successfully complete projects which previously might have been dropped.

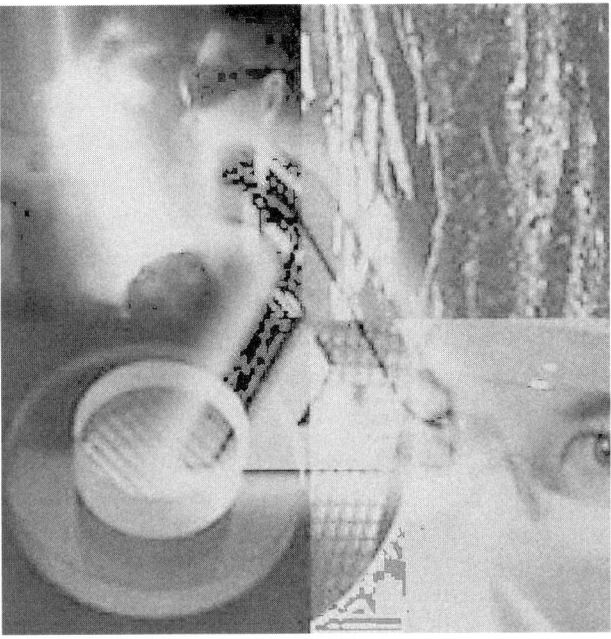

Figure 1.1 Greek letter lambda underlies the best practice in surface preparation.

1.2 Definitions of Processes Used in This Book

The following are definitions of processes:

- *Machining:* Uses fixed or rotating cutting tools to shape a part. It can cause severe damage and distortion to the surface of the material but is obviously an essential step in the process to the final surface. It generates the form or shape of the work.
- *Grinding:* Uses abrasive fixed in a fast rotating wheel. It applies lower loads to the part than machining, and so can use kinder methods of holding the part, which are less liable to cause distortion. It is often used as a second- or third-stage process in preparing surfaces.

- *Lapping:* Uses loose abrasive. It is a comparatively light load process. The pressure must be sufficient only to maintain rolling action without crushing the grit particle or jamming it into a local variation in the sample shape or hardness. This pressure is in the region of 200 g/sq cm (2.84 lb/sq in) of sample area. At higher loads, the leading edge of the sample can just wipe the abrasive out of the way, a rolling film of abrasive cannot be maintained, and severe scratching and galling results. This is often due to chips breaking away from the edge of the sample. At lower loads, the stock removal rate tends to be proportional to load. This is the largest factor in the ability of a flat lapping plate to produce a flat surface.

- *Polishing:* Normally uses a loose abrasive or one suspended in a fluid media, which may vary in its chemical properties. It usually requires a higher load, in the region of 500 g/sq cm (7.1 lb/sq in) and a softer, "mobile" polishing pad surface which can hold sharp abrasive particles so they can cut the sample. Stock removal is largely proportional to load so the plate softness introduces a major issue, that of edge rounding. Increased pressure due to distortion of the pad surface at the edge of the sample deforms the plate surface and this preferentially wears away the sample edge. As in lapping, increased load at the edge of the sample can lead to small chips breaking away and causing scratches.

- *Smoothing:* Uses a fine fixed abrasive in a rigid matrix which may be anything from a *thermosetting* plastic, to *bronze* and steel. It is often used to remove subsurface damage caused by grinding or lapping and to reduce polishing times, particularly in the optics industry.

1.3 Lapping, Grinding, and Polishing Abrasives

There are three types of abrasives which are used for surface finishing. Lapping abrasives have tough grains which roll between the sample and the comparatively hard plate surface without breaking up. Examples are *diamond (monocrystalline)*, *boron carbide*, *silicon carbide*, and *alumina*, in descending order of hardness and cost. Lapping is the process of removing material from a sample surface by rotating a hard piece of grit against it under light pressure, producing a gray finish as shown in Figure 1.2.

In grinding, the abrasive is fixed in a fast rotating wheel on a stiff machine. It is rarely used in the preparation of new materials because it causes high levels of subsurface damage, which have to be removed by lapping and polishing. However, because of high stock removal rates compared with lapping, it is often used as part of a multistep process in production environments. For optical and crystalline components, the abrasive is usually monocrystalline diamond and the process is carried out at slower speeds and renamed smoothing. Specifying abrasive wheels and blocks is discussed in Chapter 8.

Figure 1.2 These 3-inch lithium niobate wafers have the typical matte gray finish produced by lapping and are ready for polishing.

On the other hand, polishing abrasives are usually manufactured (with internal stress) to breakdown under pressure so that small, sharp, newly split particles can be embedded in the soft polishing plate surface and then cut the sample surface to produce a fine finish. Examples are diamond (*polycrystalline*), *alumina, cerium oxide,* and *colloidal silica.*

For all abrasives, it is essential that uniformity of particle size and composition is maintained. This way, larger or harder particles do not cause nonuniformity in the moving body of slurry, or "clumping" of particles, which causes similar problems. Crushed grains, or clumps of grains, can cause log jams, and scratches result. The suspension fluid can help, and although it is relatively unimportant for first experiments, both lapping and polishing can radically be affected by the chemical and electrolytic environment in the slurry. Detailed notes on specific abrasives are given in the Glossary and a comparison chart of grit sizes and uses is in Chapter 4.

A sample shape established during a machining, grinding, lapping, or smoothing sequence is relatively difficult to change during subsequent polishing.

2

Preparation: Before the Start

The number of parts to be processed determines, above all, the level of investment available for carrying out the tasks described in this book. It explains why smoothing tools are not available to speed and facilitate everything from jig maintenance to subsurface damage removal in small labs, producing a few samples for research. Equally, it explains why it can be so time-consuming and frustrating finding the image in an *autocollimator* setup, when a more expensive digital instrument, with a wide-range sighting eyepiece could save hours. Finally, it explains why, to justify the move from expensive glass optical components to injection-molded plastic ones, it requires a total volume of over 10,000 parts. Every step in a process must take account of the potential number of parts to be worked and the investment which will be required before the process route, which has been so painstakingly developed on a few samples, can be scaled up to meet anticipated volumes.

2.1 Plates and Measurement

Now to the preparation required before a real process begins. Preparation can be extensive and boring, but will reap dividends.

Most equipment found in an existing lab will require attention before being allowed anywhere near a costly sample. The sample itself will need initial measurement: dimensions, surface finish trace, and possibly photomicrographs, depending on the level of reporting required. Neglecting these steps will be a matter for regret within a day!

2.1.1 Plate Measurement

Throughout this book, reference is made to a gauge. This is a simple spherometer with three feet, on a suitable (say, 4-inch or 100-mm) pitch diameter, and a central shoe about 8-mm (0.315-inch) diameter preloaded by a mechanical or electronic dial gauge (with 2μm, (79 microinch) or better resolution). Without this, or something similar, it is almost impossible to assess the condition of either a lapping or a polishing plate.

When in use, the gauge is set to zero on a reference flat (Figure 2.1) and then placed on the clean plate surface. Slight movement will then settle the reading, allowing the convexity (center higher than edge) or concavity (center lower than edge) of the

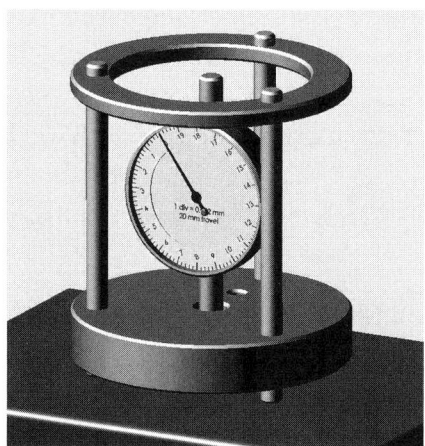

Figure 2.1 Single-dial gauge.

plate to be measured at this point. As long as the plate is clean, it can be rotated slowly while the gauge is held steady by hand, with the outer feet in contact with the plate surface. The cyclic variation in the plate shape (Section 2.4) can then be seen on the dial. The zero of the gauge should be checked on a flat reference surface plate after this, in case wear of the gauge feet has occurred.

2.1.2 Maintaining Plate Shape

Having assessed the variation in gauge reading, the plate can be conditioned to remove it (Section 2.6), so that a single reading of the gauge then indicates whether or not the plate is flat. A conditioning block, or test block (Figure 2.5) is run on the track to correct the shape to that required. This block is roughly the same diameter and usually the same material as the track width and may be up to 50% larger. Running it towards the outside of the plate makes the plate more convex. Running it towards the inside makes the plate more concave. Note that this change occurs very slowly, and times to change the plate shape by lapping are measured in hours rather than minutes. One reason why there is a recess in the middle of many plates is to make it easier to condition the plate in the concave direction.

The gauge is set up on a surface plate and zeroed by rotating the bezel. Moving the gauge slightly by hand will bed in and stabilize the reading. One small division on this dial represents the sensing tip being lower or higher than the legs by 2 μm (79 microinch).

2.2 Lapping Plates

Lapping plates must have a uniform composition and surface. Nodular graphite cast iron is the most reliable material (with glass a close second for soft samples). In this, heat treatment or additives during manufacture slow growth of the carbon particles in the iron as it solidifies from the molten state, so they are more rounded. A more uniform material is the result. This ensures that no pits or hollows develop during processing (which can result in edge rounding of smaller samples). For samples larger than 20-cm (8-inch) diameter, plates should have radial grooves to prevent build up of debris that impedes the rolling action of the grit, and then causes galling of the sample. Obviously smaller samples would catch on grooves, so these require a plain plate or mounting in groups for processing. In situations where it is difficult to apply sufficient load to the work, annular grooves may be used. Combined annular and radial grooves are sometimes used for very hard materials (Figure 2.2).

Brass (machined from a cast blank) is a good material for fine lapping plates, and glass plates are often used for lapping soft and semiconductor materials which might be contaminated by

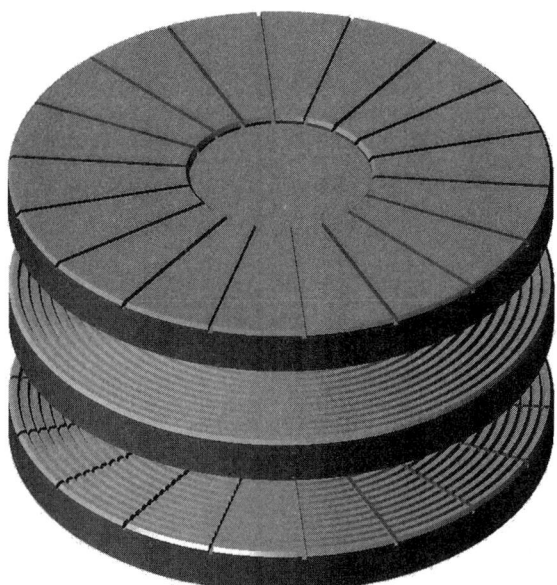

Figure 2.2 Various lapping plate groove patterns. Top—radial groove; middle—annular groove; and bottom—combined.

particles from metal plates. (Manufacture of glass plates is a very specialist job best left to the experts.)

An annular grooved brass plate some 12-mm or 0.5-inch thick is a very useful tool. With a 1-mm (0.04-in) deep recess on the rear, up to 12 mm (0.5 in) from the outer edge, creating a supporting annulus, which is lapped flat and can be bolted to the top of the normal lapping plate with a single central screw so that an extra fine-lapping stage can be done with the minimum of setup changes. The central screw allows the shape to be changed very finely, and it is easy to condition with the tools, blocks, and abrasive used for the coarse lapping process. Note that only a limited range of adjustment can be used, say, +/− 5 μm (200 micro-inch) on the gauge.

Once a lapping plate has been machined, the active surface must then be lapped flat on a larger lapping machine. If no larger machine is available, then grinding is an option. Failing this, the plate can be returned to the manufacturer for facing. (The first principal way is to have three identical plates and lap them together, by hand, in sequence. Eventually all three of them will be free from roller coaster variations and all of them flat.)

A flat plate will produce a flat sample (with some reservations). Correct sample mounting is the key. Often plates are lapped to be slightly convex using a cast iron conditioning block running on the plate itself. This can counter the natural edge rounding which occurs during polishing of a sample. Controlling and modifying the shape of the plate is followed up in Section 2.4.

2.2.1 Glazing

Some hard abrasives can cause glazing of a lapping plate surface. Here, a skin of the hard abrasive material forms on the plate and renders it useless for lapping, as the grit slides and fails to roll. The only cure for this is to lap with an even harder grit of large diameter (often for several hours). Alternatively, the top skin of the plate can be machined off in a lathe, which is more effective.

2.3 Polishing Plates

Polishing plates can have a range of surface materials, often in the form of self-adhesive pads. The key point is to have a structure which is slightly flexible or "mobile." This is fundamental and allows the surface to fix and hold particles of abrasive, which can then cut the sample.

This flexibility reflects the surface hardness which can readily be compared between plates (and the baseline, which is a granite surface plate) on a somewhat arbitrary scale. The central gauge shoe about 8-mm (0.315-inch) diameter of a standard gauge is zeroed, with the dial gauge preload only. Then the change in reading after adding a 2-kg (4.4-lb) load to the shoe is recorded (Table 2.1).

To decide on the plate hardness to use for a specific sample, the main factors are:

1. Importance of a scratch-free surface (harder gives greater risk);

Table 2.1
Typical Examples of 2-kg (4.4-lb) Load Gauge Readings for Different Plate Surfaces

Units	Microns (1 μm = 39.4 micro-inches)	Reference
Granite	0–2 (baseline)	—
Soft metal	1–3 (tin, lead)	—
FR4	4 (circuitboard material)	Chapter 9
Tufnol	6 (impregnated paper or cloth)	Chapter 9
Polytron	8 (proprietary nylon material)	Section 8.4.2
Paper (PSU)	15 (proprietary self-adhesive)	Section 8.4.2
Polyurethane	15–30 (various levels of hardness)	Section 8.4.2
Chem-cloth	50 (proprietary poromeric)	Section 8.4.2

2. Importance of edge roll-off (harder gives less roll-off);
3. Process difficulty (harder is more difficult and time consuming).

2.4 Polishing Surfaces: Care and Conditioning

Once the decision of which polishing plate to use is made, its condition needs to be assessed (if it is not possible to decide which pad material to use, then the plastic or paper-based, self-adhesive pad is a good overall compromise) [1]. Once again the gauge is the tool to use, as shown in Figure 2.3.

Run the plate under the tap and give it a light scrub with plain water; if it still looks dirty then replace the pad. Whether the pad is new or used, put the plate on the polishing machine, and rotate it at about 5 rpm, or as slowly as possible. Lower your gauge onto the rotating surface and steady it in a fixed position by hand.

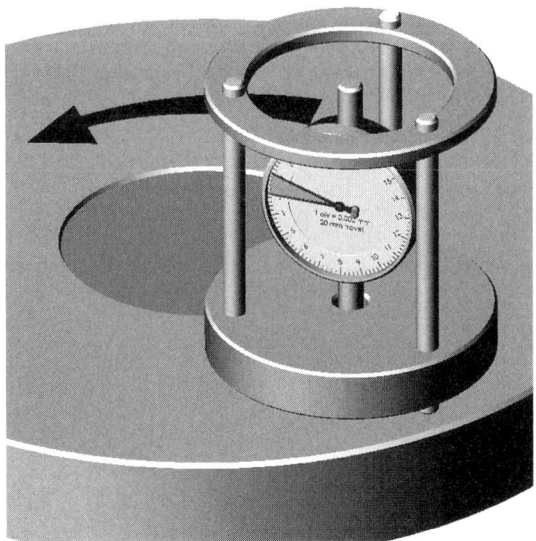

Figure 2.3 Single-dial gauge, on a plain polyurethane plate for detection of roller coaster errors.

Observe the dial. The needle is liable to sweep over anything from 2 to 100 μm (0.08 mil to 0.004 inch) as the plate rotates. This is the "rollercoaster" effect that interferes with any measurement of plate shape. Unless removed, it can make polishing much more difficult than necessary.

Even a newly fitted polishing pad will show local variations of up to 100 μm (0.004 inch). Imagine the sample negotiating these hills and valleys and the effect this will have on the polished shape. Certainly the edges will be rounded. This is why baseplates for polishing are often lapped up to 10-μm (0.04-mils) gauge-reading convex (this solution is good enough only for rough work). For demanding applications like semiconductor delayering, any roller coaster effect worse than 2 μm (0.08 mils) will make the job impossible. So will local holes in the polishing surface, which can also be detected with the gauge.

Removing worn or dirty pads may look daunting, but in fact running the plate under hot water for a few minutes will allow them to be removed easily. Time spent in measurement and preparation (lapping with a conditioning block) of the baseplate surface before fitting a new self-adhesive pad will pay dividends in time saved later.

The range of possible polishing surfaces is investigated in Section 8.2. Certain polishing surfaces, those with local variations in hardness, such as polyurethane, are difficult to assess, and this is tackled in Section 2.6.

The gauge is retained by hand, while the plate is rotating very slowly. Periodic variations in dial reading indicate the magnitude of the roller coaster effect. Local changes due to hollows in the surface become visible as sudden movements of the pointer. If a 2-kg (4.4-lb) weight is located on top of a spindle, the gauge reads minus 20 μm (−0.008 inch). The gauge anvil (8-mm or 0.315-inch) diameter has sunk into the plate surface, giving an indication of the plate surface hardness, which can be compared with other surfaces (Table 2.1).

2.5 Baseplates for Polishing

Removing roller coaster effects is not simple. Short period variations in the pad surface can be removed using a diamond conditioner on the polishing pad's surface, a job described on the next page. Long period variations (two or three peaks per revolution) are more difficult, and usually have to be tackled by correcting the baseplate itself. Sometimes a skim of the baseplate in a good lathe will correct them, but often a poor lathe will introduce a tri-lobe variation worse than the original. The safest method is to lap the baseplate on a large lapping machine (if necessary at the manufacturer) which is a complete cure. (See also Section 2.2 on lapping plate preparation). Even running a very large test block (up to three-fourths the diameter of the baseplate) will not remove long period variations, so it is worthwhile investing in spare baseplates which can be kept in good condition by sending them back to the manufacturer for lapping.

2.5.1 Baseplate Materials

Stainless steel baseplates are sometimes necessary for critical applications, particularly in the semiconductor industry. However high chromium and nickel (for example, SAE316) nonmagnetic steels tend to be unstable and must be regularly checked for distortion and relapped. On the other hand, the high *ferrite* stainless steels (which can *rust* a little and are slightly magnetic), are often a good compromise, while cast aluminum plates can be stable enough for most applications and are much lighter to handle.

Plates (including those in aluminum and brass) made from rolled plate stock are rarely stable as they tend to bend along one axis, giving a saddle shape. This is the most difficult kind of cyclic variation to cure. Cast iron baseplates can be used, and often an old, worn, and very stable cast iron plate can be pressed into years of further service as a polishing baseplate by having it epoxy painted; a very durable finish which can even be fine lapped or conditioned with a fine smoothing block (Figure 2.4) to shape the top face.

This is constructed with *bronze*-bound pellets containing a matrix of uniform size monocrystalline diamond particles, which may be from 40 to 400 μm in diameter. Its use is described in the next section. It is one of the most useful tools for keeping polishing plate surfaces in good condition.

2.6 The Use of Smoothing Blocks

The gauge indicates whether a plate surface is concave or convex, and the extent and position of any cyclic variations. A marker pen can be used to record details on the edge of the plate. Any cyclic variation with a period less than one-fourth of the plate circumference can usually be removed using a diamond conditioner as indicated in Figure 2.5. There are many types, but those which use bronze-bound pellets impregnated with diamond grit have the advantage in that they can be conditioned or sharpened.

Figure 2.4 Photograph of a diamond smoothing block.

Figure 2.5 Diamond smoothing block on an annular grooved polyurethane plate.

This is achieved simply by running the block on a lapping plate with a size of alumina grit, which is about a third of the diamond grit size. The alumina will then *undercut* the bronze to the point where blunt diamond particles fall out and the conditioner is both sharpened and shaped at the same time. Obviously the lapping plate used must itself have a shape that is inverse of the required conditioner shape.

The lubricant for use with conditioning blocks depends on the material; water for most plastic-like surfaces and nothing, or a minimal amount of oil, for paper pads. Some plate surfaces are liable to patterning during conditioner use and must be run with random radial sweep applied to the block to avoid this. The conditioner is simply run over the full track of the plate at a moderate speed until the gauge shows that unwanted variations in the plate shape have been removed.

Normally pitch, polyurethane, and cloth-based chemical polishing pads [2] cannot be absolutely measured by a gauge as the readings are confused by variations, not only in the shape but in the hardness of the surface (Table 2.1). Huge effort, however, is put into the manufacture of cloth and multicell materials so you can assume they are reasonably uniform in thickness and simply ensure the baseplate is flat or often slightly convex (see Section 8.2). The only solution for accurate testing is to run a soft glass lapped test piece on the pad for subsequent measurement by gauge or interferometer. This can also be used to check for edge rounding, which gives an indication of the severity of any roller coaster errors.

For *polyurethane* surfaces [2], coarser bronze-bound pellets (say, 200- to 400-μm (0.008-inch to 0.016-inch) diameter diamond particles) are used to remove *alkaline silica sol* deposits, but are often rather unsuccessful in altering the surface shape. This is simply because, in small labs, it is very difficult to find facilities where large particle conditioners up to 8 inches in diameter can be refurbished and shaped. A dedicated lapping plate is required, and facilities for handling large diameter alumina grit over 100 μm in size are needed. However, it is reported that when properly prepared and sharp, coarse diamond conditioners have extremely successful results in shaping polyurethane in a very short time, less than 1 minute [3]. The addition of annular grooves facilitates correcting and changing the polishing surface shape. These are machined into the pad surface on a lathe. They are approximately two-thirds the pad depth, cut with a diamond-tipped 60° included angle tool, at a pitch along the pad radius of approximately 4 mm (0.016 inch). This gives a much-reduced contact area with the sample and is a very effective way of both increasing the local load on a sample when other methods of applying load are not available, and facilitating conditioning.

If it is essential to use a polyurethane-surfaced plate for precision work, then the best method is to turn the top surface flat after sticking the pad down before using the diamond conditioner. Use a good lathe with, if possible, a diamond-tipped tool.

This type of block can not only condition a range of plate types, but can itself be "sharpened" and shaped to produce slightly different plate surface forms. Moving the diamond conditioner block in along the plate radius encourages the shape to change in the concave direction whereas moving it out causes the plate to become convex. Without a center recess in the plate, it is very difficult to make the plate shape move concave. A weight is sometimes added on top of the lifting ring on the conditioner as the concave and convex change in plate shape is very slow, usually under 1 μm (40 micro-inches) per hour.

2.7 Jigs

A jig is a device to hold the sample and present it to the polishing or lapping surface in such a way that the load and angle of the sample can be controlled. The sample can normally move up and down to accommodate roller coaster effects while accurately having its angular orientation to the jig outer ring maintained. The stock removal rate can normally be controlled by adjusting a built in load control mechanism.

2.7.1 When to Use a Jig

Polishing loads of 500 g/sq cm (7.1 lb/sq inches) would require a steel weight some 50-cm (20-inch) high if it could only be applied over the sample's own area. This is obviously impractical for, say, a 1 cm. or one-half-inch square piece. Thus, individual small samples almost always require a jig, but larger ones, or groups of samples, may sit stably enough on the polishing surface without a jig. This is providing of course that it is not necessary to control the angle of the polished surface to the sample body. Controlling this angle is a primary function of the jig.

When the sample is in a jig, which is stabilized by a heavy ring running at a larger diameter on the plate, then it becomes possible to control the angle of the sample and, for thin samples, its parallelism between the polished and back faces.

2.7.2 Jig Balance

The angle (or parallelism) of the sample can change during a long process. If the load applied to the sample, or the jig, is not central (i.e., balanced), then the angle can change due to uneven wear round the bottom face of the outer ring and the polished surface may no longer be parallel to structures in the sample body. This problem can be avoided by making sure any fixture used to support the sample does not throw the jig and its sample out of balance.

A quick check can be made by placing the jig on its side on a cleaned, level, surface plate and observing whether or not it rotates to stop in a consistent position. Balancing weight must then be added to the side of the jig which is uppermost. A typical weight can be seen in Figure 5.5, where it is sited to counterbalance the weight of the dial gauge.

Typical jigs can be seen in Figure 3.1. The white vacuum connections on top provide a low friction connection in order to retain the sample mounting plates. The gauge in this case registers material removed from the sample but also shows up cyclic variations in the plate shape. There is a knurled ring behind the dial gauge which controls the sample load. Abrasive slurry from the drum of the delivery system drips down a flexible wire onto the plate.

2.7.3 Jig Maintenance

No mechanical device is perfect, and the jig mechanism which generates the parallel motion of the sample must have some play, clearance, or compliance. This means that the sample rocks slightly and will thus process samples slightly convex on a flat plate, especially during polishing with a higher load applied.

Thus the mechanical condition of the jig is very important. If there is any play in the mechanism, it will be reflected in a convex and possibly irregular sample shape. Hence, the tendency is to make processing plates slightly convex to counteract this. (If there is consistent evidence of the jig producing an irregular sample shape and there are no loose fixings, then it must be returned to the manufacturer for repair.)

There is an exception used in geological sample lapping, where the sample is clamped solid in the jig relative to the outer ring, and the abrasive is allowed to generate whatever clearance between the sample and plate it will. It is called *undercut* and may be used to lap large thin parallel samples with little or no edge roll-off (Section 5.1). In jigs for polishing operations and light lapping, the sample mounting surface is free to move vertically and the movement constrained by very accurate mechanism to remain parallel to the outer ring surface after lapping in. For heavier lapping operations, however, such as in geology, the mechanism is clamped or constrained so that the sample is fixed vertically relative to the outer ring.

Obviously, if a thin parallel slice is to be developed by lapping, then the plate and the sample mounting surface in the jig must be flat and parallel. Fortunately, in most jigs the sample mounting surface may be lowered to touch the lapping plate so the two can be "lapped-in" together to achieve parallelism. The surface of the jig outer ring is also lapped flat in this operation, which is very important for polishing operations. Here the even rotation of the jig on the polishing plate is often a guide to how true the plate surface is, providing the jig outer ring surface is itself flat.

After lapping-in, the sample mounting surface is then raised and the sample mounted on it (or more commonly on an intermediate sample mounting plate). It can then be raised or lowered without altering its angular alignment to the plate.

2.8 Sample Mounting

The sample must be held onto the jig mounting surface so that the top (unprocessed) surface is flat against it and relatively unstressed. When released, the sample should retain the flat and parallel form that so much effort has been put into achieving during lapping and polishing operations. There are four main methods of bonding: vacuum, waxing, surface tension, and epoxy bonding.

2.8.1 Vacuum Mounting

Small components (under a 1-inch diameter) can only be wax mounted, as they have insufficient area for nature's atmospheric pressure of 14.7 psi to hold the sample in place by vacuum for processing. In fact, wafers below 75-mm (3-inch) diameter are rarely held by vacuum as they can be so thin that the grooves in a vacuum chuck will cause distortion, and thus artifacts in the polished surface, when the sample is released. Wafers over 100-mm (4-inch) diameter can often be retained by surface tension providing they are constrained at the edge (Section 2.8.4). Thus direct vacuum mounting is rarely seen. The samples are held by wax or epoxy resin onto a glass puck or metal mounting plate, which is itself held by vacuum to the mounting face of a jig. Porous sample mounting faces can help vacuum mounting, but the vacuum tends to fail during processing as abrasive finds its way under the edge of the sample.

In applications like geological thin section processing, an intermediate sample-mounting glass slide of larger area than the sample is used to ensure that no movement occurs during processing. In polishing of geological sections, a form of surface tension mount (Section 2.8.4) is used where the surface tension of the polishing fluid retains the glass slide vertically, and sideways movement is prevented by metal retaining bars at the side of the mount, which are lower than the slide thickness.

In Figure 2.6, the sample mounting face is lapped-in to the drive ring so the two are accurately coplanar. The mounting face can be raised or lowered without altering this alignment in order to accommodate the sample and mounting plate thickness. The grooves on the sample mounting face are for vacuum mounting of either a wafer-shaped sample or a separate sample mounting plate. In this case, a glass puck is held by vacuum onto the jig mounting surface. Pucks are prepared flat and parallel to within 2 μm (79 micro-inches) and facilitate removing the sample temporarily for test and evaluation.

2.8.2 Wax Mounting

For a small, rigid sample there is little choice but to wax it to the mounting face of the jig, or to an intermediate plate which is then held to this face by screws, or by vacuum. The trick is to avoid applying stress to the sample during this operation or the sample

Figure 2.6 Glass mounting puck on the jig vacuum chuck.

may change shape on release, after processing. (This includes further operations such as finishing a face with a different process: polishing a lapped face can result in a sample bending concave as the surface stress from polishing is removed. The subject is elaborated in Section 2.10.1.)

Various types and melting points of wax are available. Generally, higher temperature waxes are stronger and the danger of stressing the sample is greater. One of the best compromises is a 3:1 mixture of resin and beeswax, which melts about 70°C and is tough without being brittle. A hotplate at 75°C (which is a reasonably user-friendly temperature) can be used to bond and release the sample. However, the component waxes must be pure and free from particulates, or the bond will have grit underneath it, destroying the sample shape and alignment. To achieve this, the hot mixed wax can, with advantage, be vacuum filtered.

The question of whether to apply a load to the sample while the wax is setting is difficult and there is no straight answer, other than trial and error. This decision dictates whether or not your sample will be the same shape after processing as before. In addition, it determines whether, when released, it will have the same shape you saw when it was tested during processing. When in doubt, always view the sample surface shape before and after bonding in an interferometer (even after using one of the automatic bonding machines described later). In the author's view, applying load often causes problems, and it is better to bed the sample down into just molten wax by hand, and then leave it to relax while the wax cools. Of course, in this case it is not possible to assume that the rear face of the sample is parallel to the mounting plate and alignment should be made from the front face. Sometimes an improved bond alignment can result from laying tissue on the molten wax before bedding in the sample on top of the tissue.

2.8.3 Automated Bonding

A satisfactory stress-free bond is comparatively easy to achieve in small samples. The clean mounting plate is simply heated up to just above the wax melting point, a small amount of wax dropped onto the plate to melt, and the sample placed into the molten pool. It should then be moved around about half its dimensions to force out excess wax and then left to cool naturally. The best example is in mounting of samples for delamination of semiconductor devices, which is followed up in Chapter 9.

All waxes will contract with surface tension and this force increases rapidly with reducing gap between sample and plate. Applying local or evenly distributed pressure can result in local "suck down" of the sample area to the mounting surface and distortion, which is difficult to remove without removing and rebonding the piece. Only trial and error will prove this, and by far the most consistently successful method is the simple hand bond suggested at the start of this section. Special spring-loaded fixtures for applying load during bonding are very useful for thick samples (e.g., in geology, optics, and edge coupling work) but can apply so much localized load on thin samples that permanent distortion of the sample results.

For larger samples the story is more complex. A 50- or 75-mm (2-inch or 3-inch) diameter wafer, if bonded by this method, will end up several microns convex in because the molten wax surface tension tends to suck excess wax into the center. Pushing the wax out with anything other than a diaphragm under pressure will cause local distortion which is virtually impossible to correct. Fortunately, equipment is available (Figure 2.7) to automatically raise the temperature of the mounting plate and sample. A positive pressure is then applied above the diaphragm while the assembly cools, which counters the distortion. Evacuating the sample area under vacuum below the diaphragm further improves the bond. (Vacuum alone does not remove the minute bulge in the center of the wafer.)

2.8.4 Evaporated Wax Films

For the most precise bonding of wafer samples, the most sophisticated technique evaporates, under vacuum, an even film of wax on to the mounting surface. An automated bonding unit is then used to press the sample into this film using a diaphragm, at a temperature just below the melting point of the wax so that uniformity of the wax layer is maintained. Specialist high-temperature waxes are used which do not release large amounts of gas on melting; otherwise, evaporation of the wax film could not take place.

The unit shown in Figure 2.7 can apply pressure to the top of a wafer while simultaneously evaporating air from the bond area between wafer and mounting plate.

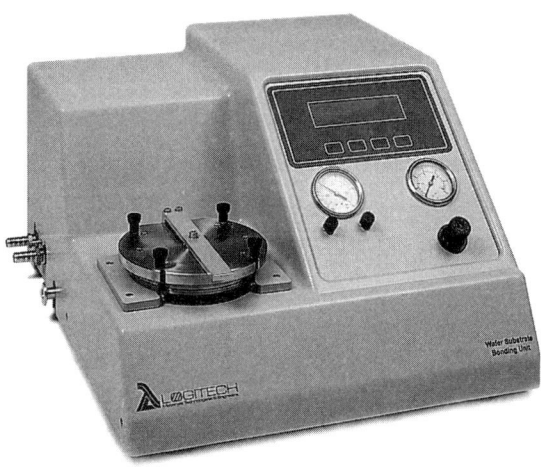

Figure 2.7   Automatic wax-bonding unit [4].

2.8.5 Surface Tension Mounting

Wafers of a 3-inch diameter and up can be held back to a mounting face by surface tension. A closed cell porous rubber pad is placed between the wafer and the mounting face and the surface tension of the polishing fluid is sufficient to hold the wafer in place, provided sideways movement of the wafer is restricted by a metal or composite ring. Proprietary rubber pads are available that provide the necessary uniformity in thickness. This method is an introduction to more sophisticated semiconductor processing equipment where the wafer carrier is controlled and loaded automatically by a column integral with the machine. The wafer floats within the carrier so that a large uniform load can be applied, which can reduce processing time to minutes or even seconds. For the first time we see real polishing loads in the region of 500 g/sq cm (7 psi) that otherwise could not be applied to wafer areas, which today can exceed 1,000 sq cm (approximately 150 sq inches). This type of wafer carrier can be seen in the machine in Figure 4.7.

Part of this reduction in process time can be due to chemical enhancement of the process, particularly with part-processed wafers, where chemo-mechanical polishing of evaporated metal layers can be the only way to restore a plane surface on partly completed semiconductor dies ready for the next evaporation layer. The machine shown also gives the opportunity to "tune" the shape of the sample mounting surface, often in real time, so that uniform removal of material across the whole area of the wafer is achieved.

The costs of line production equipment are high and there are many advantages in using smaller, sophisticated machines to test and develop processes prior to, or in parallel with, this investment (Section 4.8).

2.8.6 Epoxy Bonding

Epoxy bonding is sometimes used where the sample can be left permanently on the mounting plate, for example in thin geological section preparation. Where the sample material is able to withstand high temperatures, such as diamond or silicon carbide (these materials can require high loads during processing) then epoxy resin can also be used and removed afterwards by heating the mounting puck. This technique requires a high carbon stainless steel (SAE416) mounting puck, which is less likely to distort on heating.

2.9 Sample Viewing and Assessment

All methods of bonding rely on being able to view the results of the bonding process and ensure sample flatness and location is satisfactory before starting the processing.

Figure 2.8 Grazing incidence interferometer [5].

Conventional interferometry can be used, but obtaining a simple interferometer image of the sample surface is often impractical and very time consuming. The contour height spacing in visible light is too small (approximately 0.3 μm or 12 micro-inch) and unprocessed or lapped surfaces will not image in such an instrument. A less sensitive instrument is required, which can image lapped or unfinished samples so that the sample surface after bonding can be checked for distortion. In addition, the angle of the sample to the polishing plate must be corrected, so that the back surface remains parallel to the front during processing.

Fortunately such an instrument exists, and fast alignment and shape correction of samples are much more easily performed using a grazing incidence interferometer (Figure 2.9) with a vertical resolution of 2 μm (80 micro-inch) per fringe instead of 0.3 μm (12 micro-inch). The jig (or sample on its mounting plate) is simply placed on the flat of the instrument for the shape to be instantly displayed, whether the sample is polished or simply fine lapped. The parallelism of the bond can also be assessed if the mounting puck surface is also imaged below the sample (Section 4.7).

In the image shown in Figure 2.9, if a straight line is drawn between the intersection of the longest fringe and the edges of the sample, then the deviation from straight of this fringe (in this case, about one and one-half fringes) represents the out of flatness (i.e., about 3 μm or 12 micro-inch) of the sample. Deciding whether this deviation is concave or convex is simple. Press lightly on the sample and observe which way the fringes move. If they move towards the center of the fringe pattern, then the part is concave. If away from the center, then it is convex.

The jig is sitting directly on the reference flat with the sample suspended a few microns above it under control of the jig. Here the sample is retained on its mounting plate by vacuum during

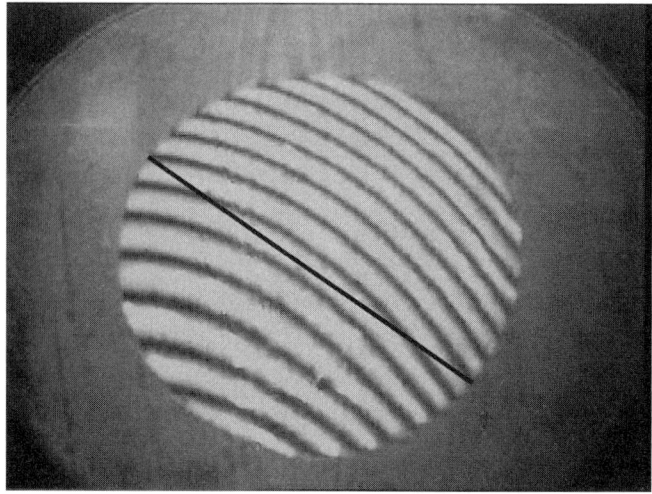

Figure 2.9 Grazing incidence interferometer image.

testing. Direct contact between sample and flat is often used to find the image. This inevitably causes damage to the flat over a time period as shown below, but instruments, even in regular use, have shown a remarkable durability of several years. A typical image is shown in Figure 4.4.

2.10 Plate and Sample Flatness Control

Often there is a situation where it seems desirable to alter the flatness of a plate to match a distorted sample. For this, there are options enabling the plate shape to be tuned. Such a plate is shown in Figure 2.10. The gauge can be used to measure the shape, but it is important to limit use of the adjustment to a very small range (e.g., +/− 5 μm, +/− 200 micro-inch), as above this, roller coaster changes in the surface can be introduced. However, it is often easier to tune the sample shape to match a flat plate simply because the tools for observing samples (interferometers and microscopes) assume starting from a flat reference.

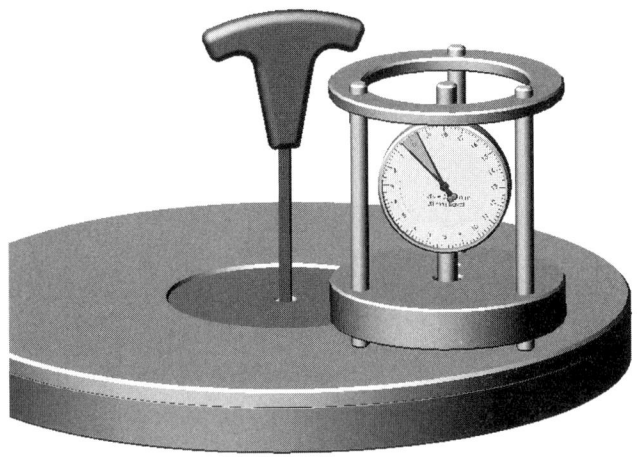

Figure 2.10 Adjustable plate showing adjustment screw.

Tools are also available to distort the sample mounting plate so that a flat sample can be achieved, both for samples which have been mounted by waxing and those held by surface tension. For waxed samples, this accentuates the need for careful bonding so that, even if the sample is concave or convex, the shape is regular (i.e., any fringe patterns seen in an interferometer are circular).

2.10.1 Wafer Distortion

Fine lapping and then polishing the cut surfaces progressively removes stress from the cut wafer surfaces, and great skill is then necessary to obtain an end result that is flat. An idea of the extent of the problem can be gained from the fact that a 1.5-mm (one-sixteenth-inch) thick slice from the boule of lithium niobate, shown in Figure 6.1, will bend in the concave direction by up to 100 μm (0.004 inch) if the wire sawn slice is subsequently polished. If the sample is waxed to a mounting plate, this concavity will not appear until the slice is released. This is one reason why wafers are often held by surface tension for chemical polishing operations (Section 2.8.4).

Here, it is often an advantage to tune the mounting plate shape to counteract the strong tendency for chemical process rates to be sample-velocity dependant, and remove more material towards the outside of the sample. This principle is used extensively in semiconductor wafer processing (Section 4.8).

A tunable wafer carrier is shown in Figure 4.9. As with baseplates, it is important to use a high carbon steel (e.g., SAE416,) or aluminum components to ensure dimensional stability of components if the chemical slurry is compatible. In this case the testing of the wafer is often done by stylus instruments, where it is possible to neglect the gross convexity or concavity of the wafer in software. This means that starting from a flat mounting plate is not necessary. Wafers for test are simply held back to a flat reference plate by a minimal vacuum. These features assist in balancing out residual stress from previous sawing and lapping operations so the final polished wafer can be made flat.

Wherever the mounting plate shape is to be tuned, then it is very important that the adjustment mechanism allows it to be distorted in such a way that the surface in contact with the sample remains spherical. Otherwise, carrying out critical polishing operations becomes impossible, as the pressure distribution due to the applied polishing load varies along the radius of the wafer.

2.11 Conclusion

This technology, although relatively simple in engineering terms, has many pitfalls, and is perhaps more suited to learning by apprenticeship than operation by inexperienced graduate-level staff. This is simply because it is necessary throughout the process to anticipate the next half a dozen moves and how they will be tackled, what equipment and materials will be required, and how much time these will take. In short, it is a skill which must be acquired over time.

References

[1] www.kemet.co.uk/product.asp?productID=1800&prodCat=Polishing.

[2] www.spartanfelt.com/polishingpads/glass_polishing_pads.htm.

[3] Karow, H. H., *Fabrication Methods for Precision Optics*, New York: John Wiley & Sons, 2004, p. 217.

[4] Logitech WSB, www.logitech.uk.com/wsb2.asp.

[5] Logitech GI20, www.logitech, uk.com/gi20.asp

3
Lapping

3.1 The Lapping Process

The sample is properly mounted and ready for removing material. If it is on a jig, it is possible to control the load and angular alignment relative to the plate. The plate has been checked and there is a good quality abrasive available in a consistent delivery system. First, wet the plate with the abrasive mix and run the jig for several minutes at about 20 rpm with the sample held at least 2 mm above the plate. Check that an even gray finish progresses across the surface of the jig outer ring and there are no scratches (Section 3.3).

The material to be removed from the sample has been calculated. Lower the sample until it touches the plate, check that the load is in the 200 g/sq cm (2.8 lb/sq in) area, zero the dial gauge on the jig, if fitted, and start the plate. Depending on the sample material and abrasive size, you can expect to remove up to 10

μm/minute (394 microinch/min), varying with the plate speed. A stable and progressive movement in the stock removal gauge is the best indicator of a successful run. If there are problems, check back on the preparation guidelines in Chapter 2, or see if the following solutions help.

3.1.1 If the Stock Removal Is Too Slow

Check the load, ensuring that the *live weight* of the jig components has been allowed for. Too heavy or too light a load can both result in slow progress. Usually both conditions also show scratching if there is a problem. The plate speed should be in the range 20 to 70 rpm given a plate size up to 600-mm (2 ft) diameter. It is not necessary to sweep the sample in a radial direction, back and forward for lapping, unless the sample is exceptionally hard (diamond or silicon carbide), in which case sweeping can extend plate life by avoiding *tracking*. This is local wear of the plate over an annulus in the center of the track which represents the area contacted directly by the sample.

3.1.2 If the Stock Removal Is Too Fast

This usually occurs when lapping a water soluble material. It is invariably accompanied by severe edge rounding and is due to the polishing fluid chemically eroding the sample (use a different fluid with much less water content, and/or a much finer abrasive).

This is carrying 3 jigs and a plate flatness monitor for simultaneously processing a large number of samples. The monitor is part of a control system, which moves the jig control arms to make the jigs operate towards the center of the plate, making the plate wear concave, and vice versa. It enables long processes to be carried out without constant checking of the samples and the plate shape.

3.2 Plate Shape Monitoring

Removing or avoiding the tracking effect is very important, as with tracking it becomes virtually impossible to assess the shape of the plate using the gauge on the plate itself. The gauge will of course detect the tracking as sudden changes in reading as it is moved along a plate radius. The solution is to run a test block (Figure 3.2) of the plate material either at the same time as the sample or frequently between sample runs so that the tracking does not develop. This block is then easy to clean, and using the gauge, you can measure accurately the inverse shape of the plate (i.e., the plate shape which the sample sees).

The additional wear of the test block on the plate (providing it is large enough in diameter to span the active plate track) will help to eliminate or avoid tracking. (Some machines also use a ring conditioner that achieves the same, but of course cannot be measured. In this case, a test block of suitable diameter to suit the gauge must be run inside one of the rings.) The block shape as measured then tells you what to do to maintain the desired overall plate shape, be it flat or slightly convex. The universal rule is to move the block out if you want the plate to become more convex, and move the block in if you want it to become more concave. In the concave case, the job is much more difficult and this is why most lapping plates have a hollow in the center, leaving a working annulus or track around the outside, which has a radial width slightly smaller that the jig outer ring. If tracking is severe then the plate surface will have to be turned flat on a lathe and relapped. The sweep of the jig or sample is, of course, the main weapon to reduce tracking, although it is not always possible to sweep the sample over the whole track unless it is mounted off center towards the inner diameter of the jig outer ring.

Automatic monitoring and control of plate flatness is available on some machines and is invaluable where many samples are to be processed. The plate monitoring system in the machine shown in Figure 3.1 detects tracking wear.

3.3 Scratching

Scratching is an occupational hazard. Look at a scratch under the microscope, and 99% of the time it will extend inward from the edge (Figure 3.4). A thicker slurry mixture which lifts the sample higher off the plate can help. The standard mix is 250 cc (8.5 oz) of abrasive to 1 liter of fluid. Alternatively, try sprinkling a few grains of neat abrasive onto the plate in front of the jig towards the end of the process.

Nonmagnetic stainless steel grades (e.g., SAE316) are particularly prone to this. The use of a lathe-cutting additive in the fluid can help. Also, it is worth trying the use of alumina as an abrasive rather than silicon carbide. It is particularly important here that

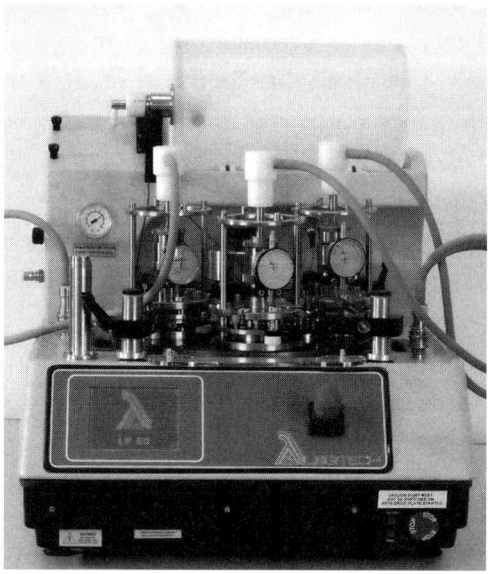

Figure 3.1 Professional lapping machine [1].

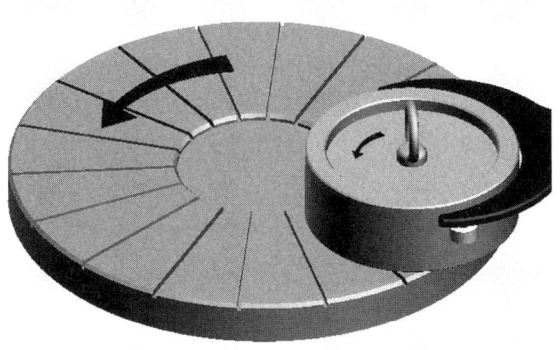

Figure 3.2 Test block on a radial grooved lapping plate.

Figure 3.3 Inverted test block with gauge in position.

there are radial grooves in the lapping plate (Figure 3.2), which are at least as deep as they are wide (say 3 mm) and not clogged with debris. A coarser abrasive, say 30 μm (1.20 mil), or even 50 μm (2 mil) diameter may also help.

Figure 3.4 Microscope photograph of a lapped surface with scratch clearly visible [×200: the diagonal represents approximately one-half millimeter (0.02 inch)].

3.4 Smoothing

Over many years in sample processing using simple lapping and polishing, there was an obvious need for a process halfway between the two: simply to remove subsurface damage, shorten polishing times, and improve the final surface form. Thankfully one has emerged during the last decade, using techniques borrowed from the optics industry where *smoothing* has been used since the 1950s (Chapter 8).

Carried out on any advanced machine, the soft nap polishing pad is replaced by a plastic/resin-fine reticulated self-adhesive pad where the top surface, up to some 1-mm (0.04-in) deep, is impregnated with fine diamond particles. These can be specified but experience with 3-μm (120-microinch) polycrystalline particles, suggests that high stock removal rates and finishes halfway between fine lapping and polished can be realized.

Wafer samples are surface tension mounted in the carrier and loads can be applied nearer to polishing than lapping (Section

1.3). The reticulated plate surface can be conditioned and shaped by running a test block in the carrier, or a separate control arm with fine alumina abrasive (about one-third the size of the diamond particles). The chemical polishing machine used can be flushed down with water and scrubbed to clean the surface before use.

The finish obtained can rapidly be polished to remove ultrafine scratches from the diamond particles in approximately one-fourth the time it takes to polish a fine lapped finish. It is worth comparing this concept with the optical practice described in Section 8.6, where it is normal to apply two smoothing stages to reduce polishing times.

In sample rather than *substrate* processing, there is a complication; that of cross-contamination of materials where the smoothing block can carry foreign particles from one type of material to another. The simplest solution is to use the bronze bound pellet type similar to the diamond conditioning block shown in Figure 2.5. This has up to 3 mm (1/8 inch) of usable depth of diamond matrix, and if the pellets are mounted on an aluminum baseplate, the active plate surface can be "sharpened" regularly between samples or batches, trimming the shape and simultaneously removing any possibility of material from one sample being transferred to another (see Section 2.6). A small-enough smoothing block diameter can be used to allow ultrasonic cleaning of the block after this, for added certainty.

A somewhat costly alternative is to use self-adhesive pads carrying blisters of resin- or metal-bound diamond grit, which can be changed between sample types [2]. These may be conditioned in the same way, a limited number of times.

References

[1] Logitech LP50, www.logitech.uk.com/lp50.asp.

[2] www.struers.com/resources/elements/12/144043/ConsumablesCatalogue2011_udenHT.pdf, p. 26.

4

Polishing

4.1 Introduction

Polishing removes very little material. The stock removal rate is usually a few microns per hour instead of per minute. Trying to remove three times the lapping grit diameter during polishing, in order to eliminate subsurface damage, while maintaining both flatness and freedom from edge rounding, is a tall order. The chances are much improved if fine lapping or smoothing has been done as an intermediate process. However, once there is an accurately flat and correctly aligned, lapped surface, polishing can be attempted.

How large does a grit particle have to be before, in hitting the edge of the sample, it chips off a larger piece from this edge and causes a massive scratch? Is this sample chip part of a single crystal material, so that when it rotates it presents a harder axis to the sample, which can then cut the surface to cause an even larger scratch? (See Figures 4.1 and 4.2.) Initial success in producing a

58 Substrate Surface Preparation Handbook

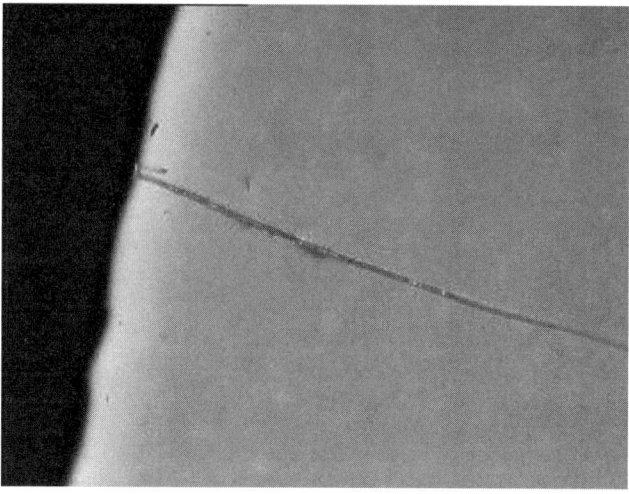

Figure 4.1 Microscope photograph of scratch on a polished surface (40×). This sample will have to be relapped.

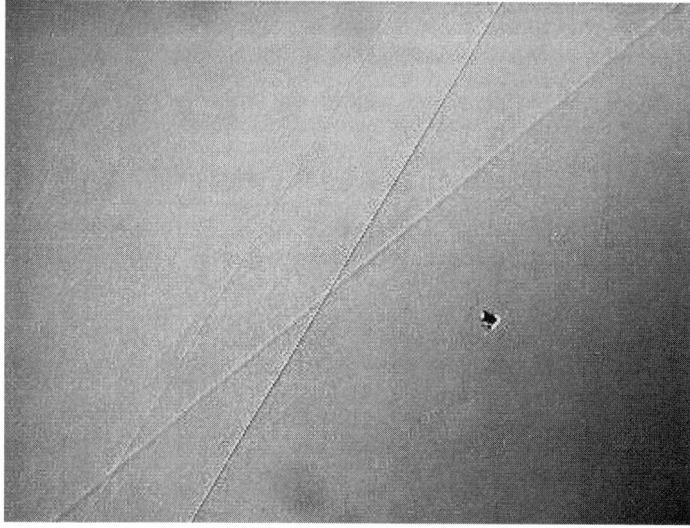

Figure 4.2 Scratches and voids (100×).

scratch-free polished finish on any combination of material, plate and abrasive will depend on all these factors. The longer the polish, the greater the edge rounding and the less likelihood of this happening; however, the worse the sample shape. If the shape is critical, it may often be worthwhile relapping or smoothing the sample to a finer finish, so that the polishing time is reduced. More important, lapping to a smaller grit size involves less material removal during polishing if this is a subsurface damage critical application. If edge rounding on the finished sample is a serious issue, then it is possible to finish the lapping plate slightly convex so the sample surface is slightly concave to facilitate subsequent polishing.

These are several reasons why a ten-minute process can turn into ten hours, or even days! The saving grace is that problems usually reveal themselves quickly and can often be rectified, providing the sample surface is checked frequently during the processes (Section 4.7).

The application is always important, as subsurface effects may or may not be critical. One theory suggests that a thin layer of sample material is smoothed over whatever damage is below the surface (the *Beilby* layer) and can be etched away to reveal the damage again after polishing. This is one reason for the popularity of chemical (Section 4.8) and chemomechanical (Section 4.9) polishing, as the formation of such a layer is inhibited by the chemical action.

4.2 Sample Load

The optimum load has been given as about 500 g/sq cm. This is a lot (about half a ton) if you have a sample of, say, 12-inch diameter. The "live" weight (weight of the moving parts) of a jig is often in the region of 2.5 kg (5.5 lb) so this can be used to load a sample up to about 5 cm² in area. Additional weights up to a further 5 kg (11 lb) can often be applied, so the maximum sample area could be 15 cm², or about 4-cm diameter. In fact, with sample sizes in this range, the full load is rarely applied, and a 5-in jig can effectively polish a sample up to 3-in diameter, especially when annular grooved plate surfaces are used. Purpose-built load gauges simply measure the total load on the sample generated by a jig or by the sample weight, if the sample is large enough to run stably (Section 2.7.1).

- *Vacuum mounting blocks:* Shown in Figure 4.3, large wafer samples can be held by vacuum onto the grooved surface of a mounting block and the adjustable load weights enable wafers up to 150-mm (6-in) diameter to be polished on standard 300-mm (12-in) plate polishing machines. At

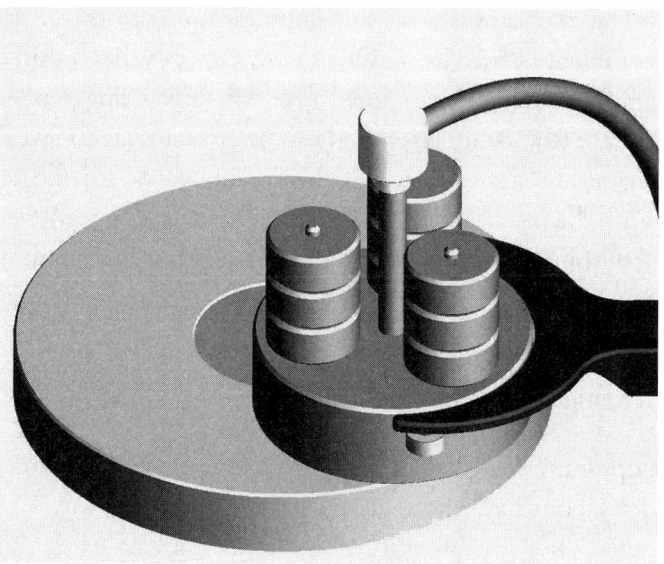

Figure 4.3 Vacuum mounting block.

this size it is usually unnecessary to use a jig, as the sample area is sufficient for the wafer to ride stably (remaining parallel) on the polishing plate. These factors effectively define the optimum size of jigs, with larger ones being used only for very specific applications, such as large geological samples. Note however, that too small a load often leads to scratching due to there being insufficient load to prevent rogue particles getting under the sample.

- *Sweep:* As opposed to lapping, with polishing it is almost always necessary to move the sample along a plate radius during processing. Without this, the polished surface tends to develop regular waves or artifacts because it is running regularly over the same part of the plate. A randomized radial movement (sweep) is easy to impart by rotating or moving the arm which holds the jig or sample mount at a speed that is independent of the plate rotation speed.

4.3 Abrasives

In polishing, the relationship between surface finish and abrasive grit size is not straightforward as in lapping. Because polishing grit is designed to break down into smaller particles, the finish that a given grit size will produce depends on a range of factors. These are tied up with the hardness and brittleness of both abrasive and sample, together with the plate surface hardness (Table 2.1). In general, it can be said that the larger the grit, the worse the finish and the faster the stock removal, but the brittleness and crystal structure of the individual particles also come into play.

Each of the abrasive materials referred to earlier have different characteristics and some will produce a good finish in certain situations, while an alternative material at the same particle size will produce scratches. A significant scratch can be very difficult to remove by further polishing and so relapping with a finer grit may be the only answer.

4.4 Edge Polishing

Some samples can be edge polished to help prevent scratching. This is quite difficult to do and is not always effective. An initial surface polish on a very soft pad (such as those used for chemical polishing) is often much simpler and often more effective. The fine, edge-rounded shape it produces can prevent particle break away, and if the gross shape is affected, this can be corrected by using a harder plate later in the process.

4.5 Slurry Flow Rate

The rate of slurry flow is significant. Often the best polishing runs are with the plate almost dry, but evenly moist. Quite apart from the abrasive cost, this is why it is so important to have good control over the flow, and constant mixing of abrasive and fluid.

4.6 Grit Sizes

Methods of specifying abrasive grit sizes are as varied as the different countries and industries which use them, and information given in Table 4.1 must be taken as approximate. Recommendations are a starting point only!

4.7 Aligning the Sample

If a jig is used, the sample surface will have to be aligned coplanar with the outer ring. The grazing incidence interferometer is very useful for this, (Section 11.3) as it can also check the sample shape, and the sample image is easy to find. When a jig is on the

Table 4.1
Grit Size and Use Comparison

Lapping Grit	Grit Material and European Size Code	U.S. ANSI Size Code (approximately)	Size in Microns
Hard materials	Diamond monocrystalline	1,500–180	3 – 80
Rock	Boron carbide F600–F400	900– 600	9–17
	Silicon carbide F600–F400d	900– 600	9–17
	Monocrystalline	1,500	3
Silicon carbide	Diamond monocrystalline,	320–180	30–80
	Boron carbide F320–F240	320–180	30–80
	Alumina F	320	30
Silicon	Alumina F400–F320	600–320	17 – 30
Other semiconductor	Alumina F600	900	9
Glasses	Alumina F800 – F600	1,000–F320	5–30
	Silicon carbide F600	900	9
Polishing grit			
Hard materials	Diamond polycrystalline	1,500–900	3 – 9
Rock	Diamond polycrystalline	1,500	3
Silicon	Colloidal silica	—	1/20
Other semiconductor	Colloidal silica	—	1/20
	Ultrafine alumina	—	0.5–3
Glasses	Cerium oxide	—	0.5–3
	Ultrafine alumina	—	0.5–3

interferometer reference flat, and the outer ring image is seen to be in intimate contact with it (i.e., the fringes are "fluffed out"), then the image of the mounting surface behind the sample shows its angle to the flat as illustrated in Figure 4.4. The number of visible fringes on the sample mounting plate accurately portrays the angle of the face, at least for thin samples (up to 1-mm thick). Recording (capturing) the fringe pattern image at the start of the process allows comparison with later images to see if anything has moved during processing. If the sample is large enough, then the fringe pattern on it can be used for alignment. The jig angular adjustment can easily reduce the number of fringes until only one fluffed out dark patch can be seen (or more likely, one or two concentric rings centered on the sample itself). If there is need

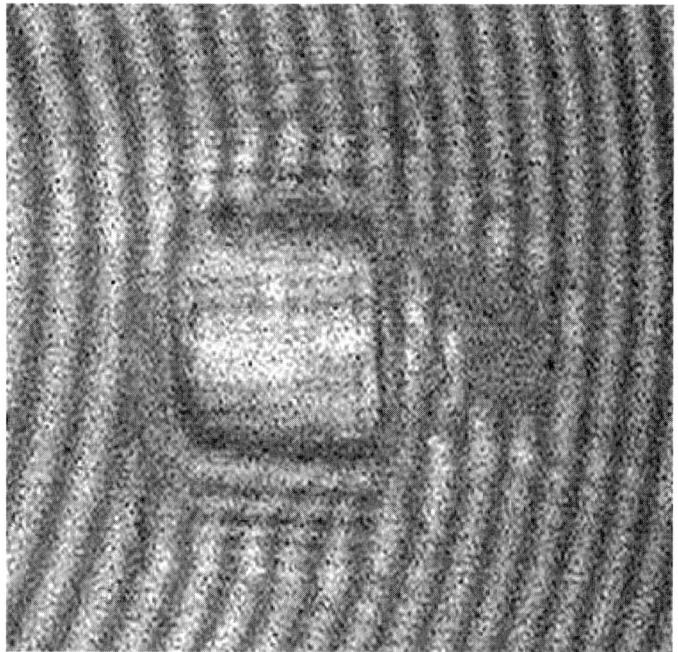

Figure 4.4 Typical fringe pattern of sample on a mounting plate. The sample is 6-mm (0.24-inch) square.

to remove material from a specific area of the sample, then this is where the rings, or fluffed-out patch, should be centered.

Obviously if there are several fringes on the sample using this instrument [at a vertical spacing of 2 μm (79 microinches) per fringe], then the sample is either not anywhere near flat or the angle at which it is mounted is incorrect.

For more accurate work the same principle at 6 times improved accuracy [fringe vertical spacing at 0.3 μm (12 microinches) rather than 2 μm above the flat] applies when the drive ring is viewed using a monochromatic light (Figure 4.5).

Caution is necessary, because the image can be very difficult to find with the flat simply laid on the clean surface of the jig outer ring. Using the grazing incidence interferometer first, can

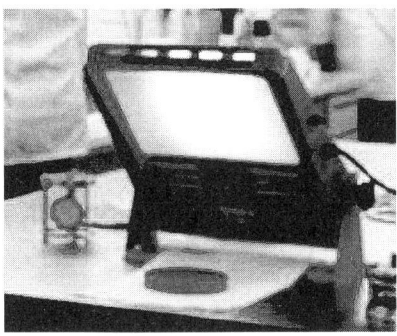

Figure 4.5 Monochromatic (sodium wavelength) light. Also visible is the 127-mm (5-inch) diameter optical flat, for higher accuracy flatness measurement and sample alignment.

enable the image to be found more easily and brought to approximate alignment.

This pattern is shown here on a grazing incidence interferometer during sample alignment. Note the relevant sets of fringes. The pattern of the sample mounting plate round the outside is some 2-μm convex. In the center of this, the pattern of the square sample itself is almost flat along the horizontal axis but along the vertical axis, a dark fringe is showing top and bottom so it is about 2-μm convex along this axis.

The slightly lighter area in between, is from the bonding wax film. There are some 4 fringes showing on the mounting plate in the area under the sample, which means that either the sample is out of parallel by 8 μm or the bond between sample and the mounting plate is wedged by the same amount. This will be clear, only if images before and after bonding the sample were captured and recorded. (See Section 2.8.2.)

This image will be sufficient to place the work in the correct plane, but after an initial run it is unlikely this will have been sufficiently accurate for final alignment. A procedure for more accurate sample setup is shown in Section 4.7.

4.8 The Polishing Run

4.8.1 Before the Start

First anticipate all the problems, and then prepare the plate, jig, abrasive, and sample: all is ready. Then thoroughly wet the plate with the abrasive slurry. The best way is to put the jig on the polishing plate, with the sample riding well above the plate (by reducing the jig load control to zero), and allow the jig to run so you can observe its movement with the abrasive feed running. After wetting the plate, gradually reduce the slurry flow until the jig runs at an even speed with little evidence of fluid buildup on the plate.

If this is the first time for this sample and plate combination, a lot can be learned from the stock removal gauge on the jig. If the gauge fluctuates wildly or if the stock removal rate is significant, immediately stop the plate. Even after a few seconds, it can be seen where the plate is contacting the sample from the shiny part of the lapped surface. Obviously if this is at one side, then the sample alignment is wrong and this could cause fast stock removal. Wild fluctuations usually occur when the sample is not fully contacting the plate.

4.8.2 Monitoring Progress

If preparations are good, the sample will drop onto the plate when released (from a few microns above it), and the stock removal gauge will hardly move during plate rotation. The stock removal gauge will show the material removed and the process can be stopped at an appropriate moment, either to check progress, make adjustments, or to conclude the process. If only a few microns are to be removed, then it is likely that cyclic variations in the plate shape will soon become larger than the remaining material to be removed; this is the time to consider reconditioning the plate surface. Check the sample surface, and if the shape has deteriorated, although the surface finish is good, consider relapping or smoothing it at the same time as reconditioning the plate surface. The next run should then produce a much better result. It is always a good idea to check the sample on any new run after a few seconds, especially if the sample movement is unstable. There are a few hints on basic microscopy in Section 10.3.

4.9 Jig Rotation

The best guide, after the stock removal gauge, to the likely success of the run, is the uniformity of the jig or sample rotation. If plate and jig are in good condition as described in Section 4.3, and the plate is evenly moist, then the jig will rotate evenly, subject only to slight periodic variation during radial sweep. Whether or not the sample is in the center of the jig will affect this. It should almost always be central. The exception is in the most tricky of operations, such as semiconductor device delayering, when an off-center location is often used to indicate just when the sample starts to contact the plate. However in this case, the run may consist of no more than a few plate rotations. The same applies to a very hard sample which must sweep over the whole track to prevent *tracking* or uneven pad wear.

4.10 Sample Surface Shape and In-Process Alignment

As soon as the surface of the sample starts to look shiny, you can view its shape in an interferometer or use a monochromatic light with a simple optical flat. This is facilitated if you are using a jig, as the flat can simply be laid on the outer ring of the inverted jig and the light shone down on the flat from above. With some samples, the alignment process (Section 4.4) may have to be repeated if the fringe pattern is either not fluffed out or not central. This is likely on the first run after lapping if the sample has initially been aligned on a grazing incidence interferometer. The improved resolution (0.3-μm fringe spacing instead of 2 μm) gives the possibility of achieving higher accuracy of alignment.

Fairly typical results are illustrated in Figure 4.6. Features on the jig allow the sample to be raised under the flat until it is a few microns clear to show the optimum fringe contrast. Angular corrections are simple to make on the jig. However, if this angles the sample so it is outside its parallelism specification, then repolishing will be required (Figure 4.5 shows this scenario at 2 μm per fringe accuracy with a lapped mounting plate). Note that at this higher accuracy, a lapped surface will not image and the mounting face will have to be part polished to show the alignment and the need for this will have to be anticipated early in the process.

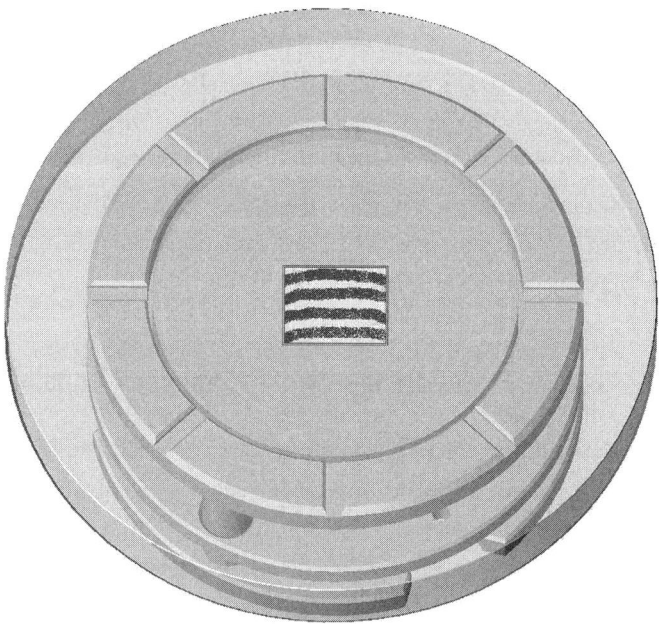

Figure 4.6 Inverted jig with 178-mm (7-inch) diameter optical flat on the outer ring.

The optical flat is resting on the drive ring of the jig, all underneath the monochromatic light (Figure 4.5). No fringes are showing on the outer ring surface, as this is effectively 100% in contact with the flat. (Various factors can, however, make this ring wear unevenly and cause fringes to appear on its contact surface.)

1. There are four fringes showing on the rectangular sample itself, meaning that the sample top edge is $4 \times 0.3 = 1.2$ μm nearer the flat than the lower edge. The jig angular adjustment can be used to correct this, and then only one approximately circular fringe will appear.
2. The bend in the fringes is approximately one-fourth the distance between the fringes. This means that the sample is 0.3/4 or almost 0.1 μm concave or convex (more often convex due to edge rounding).

Note that as the monochromatic light output is not accurately collimated, then the fringe contrast will be low. It is often necessary to put out ambient lights for this work in order to locate the fringe pattern. Use the jig sample height adjustment to bring the sample up to be a few microns only clear of the flat. In this setup, caution is necessary to ignore moving circular fringes (*Hadinger fringes*).

4.11 Chemical Polishing

If it proves impossible to achieve both the finish and stock removal required on a new sample, the answer is often found during a trip to the chemistry department of your company or local university, to find out which chemical feature may be used to enhance the process.

A good example is to try and polish copper conventionally, which is a very difficult task, even with *cerium oxide* abrasive, which is slightly acidic. Add 10% of concentrated nitric acid (which is itself 60% concentration by volume) to the slurry (observing the precautions the chemist will have told you about) and miraculously both finish and stock removal are there at your fingertips. Some materials respond to this more acidic mix and some (like gallium arsenide) to one more alkaline. The only problem is that the equipment you have must withstand this environment. In many cases, it is worthwhile investing in a chemically resistant kit just in case this situation arises. An alternative and simple example is silicon nitride, normally tackled only with diamond abrasive. However, one of the softer abrasives which is very friable, *cerium oxide*, will remove material as, in water, it is slightly acidic.

For situations where sample edge breakaway is causing scratches, chemical enhancement can be a boon. Often chemical action is itself accelerated at sharp areas round the edge and, coupled with the use of a soft pad on the polishing plate, this can often avoid scratching altogether. In semiconductor device substrate preparation it is almost always used, and the consequent edge rounding of the wafer material is accepted and allowed for (see Figure 4.10). Even the softer grade of closed cell polyurethane is often too hard for chemical polishing and soft proprietary cloth pads are preferred (Sections 2.3 and 8.2).

One of the most useful etches, particularly for gallium arsenide and other softer crystalline compounds in this group of materials, is bromine, diluted with methanol. It can be used, without any abrasive, in machines constructed within fume enclosures. The material is simply dissolved by the etch and the surface form controlled by the plate; those areas which are under higher pressure or moving at higher velocity are subject to greater stock removal. This leads to greater edge rounding and much experimentation with the shape of the baseplate (which can be a lift-off plate in glass, which gives the necessary stability and ease of cleaning) results in a surface which is slightly convex. Other parameters such as load, sample mounting surface shape, and the relative velocity of sample and plate, are then used to control the final shape of the wafer. The latest machines are able to vary these during a process (see also Section 2.8.3). More recently, bromine has largely been replaced by hydrogen peroxide in an alkaline solution for this task [1].

In this way, surfaces free from the subsurface damage (which can be caused by abrasives), suitable for subsequent epitaxial growth of semiconductor device structures, can be created.

In Figure 4.9, the lower wafer is the original with a protective coating intact. The upper wafer has been polished to remove the coating and reveal the next layer. Various hardware features were necessary to achieve this, including adjustment of the mounting surface shape so that the metal layer, some tenths of a micron thick, could be evenly removed. The polyurethane surface shown in Figure 4.8 would be too hard for this and a soft nap cloth would be preferred [2].

Figure 4.7 Chemical polishing machine showing surface-tension mount carrier with tunable wafer mounting face shape and 400-mm (16-inch) diameter poromeric polishing pad.

72 Substrate Surface Preparation Handbook

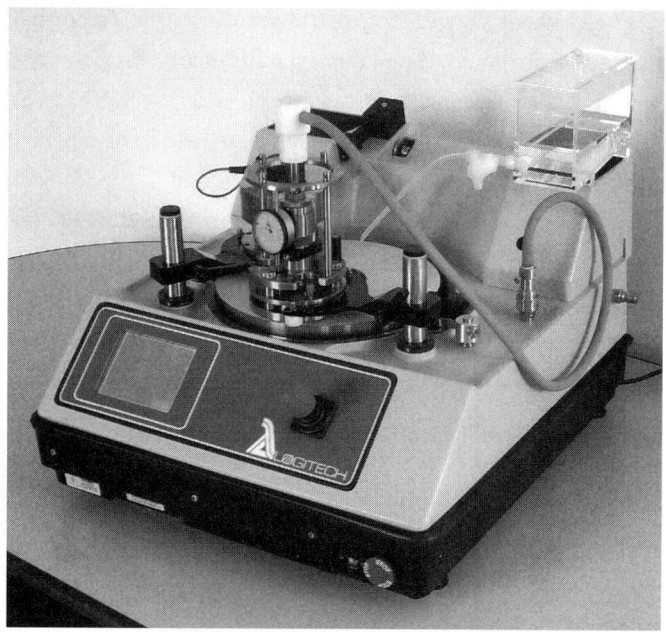

Figure 4.8 Chemical polishing using a drip-fed etch on a 300-mm (12-inch) diameter polyurethane plate.

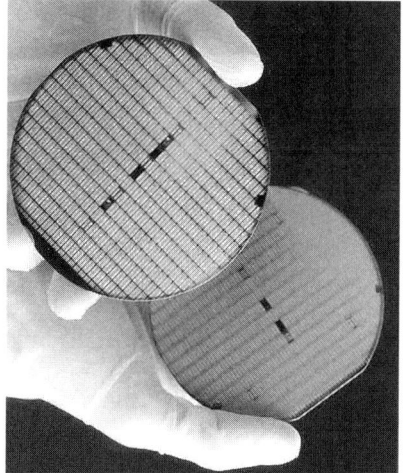

Figure 4.9 Two wafers.

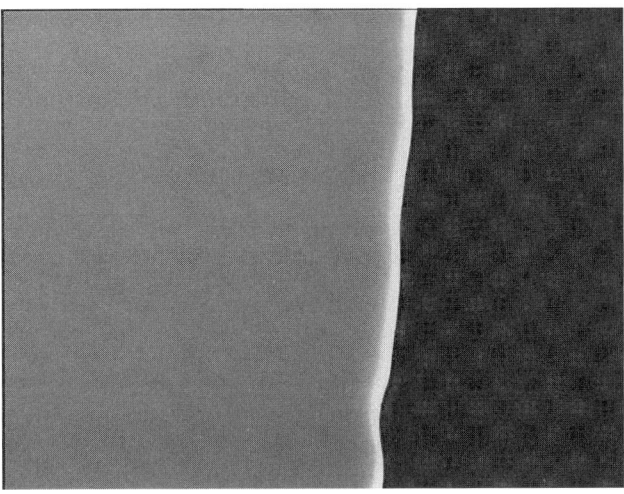

Figure 4.10 A chemically polished surface. Approximately 100 magnification (1-mm or 0.04-inch diagonal).

4.12 Chemomechanical Polishing

Considering the options of abrasive available (Table 4.1), and multiplying these by a range of chemical options, it becomes apparent that chemomechanical polishing can be a complex subject, with proprietary abrasive and chemical mixes varying with application and company. Any combination of abrasive and chemical polishing can be classed as Chemo Mechanical Polishing. It came into prominence after 1990 with the adoption of copper rather than aluminum semiconductor structures, when it was used to finish copper conductive layers coplanar with insulating *silicon oxide* or *silicon nitride* layers during the device build process.

One of the simplest CMP methods uses material that came from the building industry, where it is used for hardening concrete. Colloidal silica is an example of an alkaline silica sol: a milky liquid with a soapy feel containing minute (1/20-micron) -size particles of silica in suspension. It has a high (alkaline) pH and is very effective in polishing many metals and crystalline compounds such as gallium arsenide. It is commonly used with a soft cloth pad or with polyurethane when it forms a popular combination, particularly suited to polishing lithium niobate and other glassy materials. Work at Glasgow University has shown that modification of this material with hydrofluoric acid can significantly improve its performance [3].

Polyurethane plates normally require high polishing pressures. The resilient nature of the pad leads to significant edge rounding, and this is responsible for its reputation as a "rough" method of obtaining a quick polish. Also, the surface cannot be accurately measured by a gauge because of this flexibility, and it is difficult to condition a polyurethane pad to control the sample shape (Section 2.6). The colloidal silica is usually dispensed by a drip feed, as shown in Figure 4.7, and may be diluted with *de-ionized water* up to 4 times. A consistent delivery system is necessary with a special regulating valve which resists the high alkalinity and does not clog.

Load can be increased by grooving the pad surface so that only a narrow land every 3 or 4 mm along the radius of the plate contacts the sample. Using a good lathe is necessary, and the grooves form best when using a 60° diamond-tipped tool. After grooving, the tops of the lands should be skimmed with the diamond tool, making sure that the surface is flat. Some labs use exclusively this type of pad for general work and move to a hard alternative (e.g., soft metal or composite plate with polycrystalline diamond abrasive) only when edge rounding is a problem.

In daily use, polyurethane plates are not normally allowed to dry out overnight, as crystalline silica deposits can build up. A diamond conditioner will remove these and refurbish the surface, but usually fails to alter the plate shape unless really coarse conditioners (200- to 400-μm or 8- to 16-mil grit size) are available in good condition.

For silicon wafer polishing with colloidal silica, a *poromeric* synthetic pad is a good starting process.

From being a building material, colloidal silica has become a prime example of chemomechanical polishing processes where combined chemical and mechanical action contribute to the final polished surface.

4.13 Fluid Jet Polishing: Future Developments

New techniques lead on to further advances, and the ability to generate three-dimensional surface scans (Section 8.4) from interferometer images leads to the ability to control point-polishing techniques with sufficient accuracy to remove local variations and surface faults in samples with previously unattainable accuracy.

A case in point is fluid jet polishing, where techniques developed for water jet cutting have been adapted for polishing [4]. Many papers have been published over the last 10 years on what are relatively simple parametric investigations into this process and development has led to companies claiming exceptional results [5–7].

In addition to development of existing technologies, the exploitation of new techniques, particularly those which are software related, is crucial to maintaining the progress in this field, which has led to polishing becoming mainstream in semiconductor device manufacture.

References

[1] McGhee, L., et al., "Chemomechanical Polishing of Gallium Arsenide and Cadmium Telluride to Subnanometre Surface Finish. Evaluation of the Action and Effectiveness of Hydrogen Peroxide, Sodium Hypochlorite and Dibromine as Reagents," *J. Mater. Chem.*, Vol. 4, No. 29, 1994.

[2] www.kemet.co.uk/product.asp?productID=1800&prodCat_FK= Polishing.

[3] Beveridge, M., et al., "Chemomechanical Polishing of Lithium Niobate Using Alkaline Silica Sol and Alkaline Silica Sol Modified with Hydrogendifluoride Anion," *J. Mater. Chem.*, Vol. 4, 1994, pp. 119–124.

[4] Liu, H., and J. Wang, "Abrasive Liquid Jet as a Flexible Polishing Tool," *Int. J. Materials and Product Technology*, Vol. 31, No. 1, 2008.

[5] Booij, S. M., H. V. Brug, J. M. Braat, "Nanometer Deep Shaping with Fluid Jet Polishing," *Proc. SPIE* 4451, August 2001, pp. 222–232.

[6] Messelink, W. A., et al., "Optimization of Fluid Jet Polishing CNC Tool Design," Fisba Optik AG.

[7] www.lightmachinery.com/Fluid-Jet-Polishing.html.

5

Specific Processes and Materials

In this book, it is impossible to ignore the wealth of existing applications and their history. Up until the 1950s, most polishing work was on glass to create components for optical devices designed to work at normal visible wavelengths. With the invention of the laser in 1960, there suddenly was a demand for materials which could operate (remain transparent) in wavelengths much longer than the normal visible range, up into the infrared region, and later down into the ultraviolet region.

Mostly crystalline, there is a wide range of materials which exhibit electro-optic effects. Many are also transparent over an extended range of wavelengths and can be used as window and lens materials, as well as light modulating components in both discrete and embedded devices.

5.1 Geology

Examination of rock structures involves microscope examination of mineral crystals under transmitted polarized light, where the birefringence colors revealed as the light passes through the material enable identification of the specific minerals in the structure. For this to be successful, the rock section has to be some 27 to 30 μm (1–1.2 mils) thick. Typical sections are permanently mounted for support on glass slides, varying in size from 25-mm (1-inch) diameter up to 150 × 100 mm (6 × 4 inches). This very specific method of analysis separates geological from metallurgical processing (covered in Chapter 10), in spite of many similarities.

Special epoxy resin with the same refractive index as glass is necessary for the mounting operation and proprietary methods of bonding are used, so that any excess resin between slide and sample is removed.

The use of hard-surfaced outer rings on the jigs and lapping of both slide surface and rock sample using the *undercut* method ensures that the surfaces are extremely flat before bonding to maintain uniformity of the final sample. This is an amazing achievement considering the sliver of rock is half the thickness of a sheet of paper. The thickness uniformity maintains the integrity of the information gained during analysis of the viewed slide, such as that shown in Figure 5.1.

Figure 5.2 shows how the grazing incidence interferometer can assist in maintaining this by providing a simple in-process check to eliminate distorted glass slides. This can eliminate the several microns local variation in slide thickness which is common and very difficult to pick up by simple micrometer measurement. After rock samples are cut to size on a diamond saw (Figure 6.2), they are bonded to prepared glass slides and cut to a suitable thickness for lapping using a similar saw adapted to take several slides at a time as shown in the illustration. The lapping process varies with the material, but usually reduces the thickness of the rock on the slide to the correct thickness for lapping in a single pass. In most cases, the lapped sample can then be viewed directly after cleaning, and analysis made. This is either visual or computer aided.

Specific Processes and Materials 79

Figure 5.1 Geological thin section set up for analysis. This particular material is Scottish granite, viewed with polarized light under approximately 200× magnification (1/2-mm or 0.020-inch diagonal).

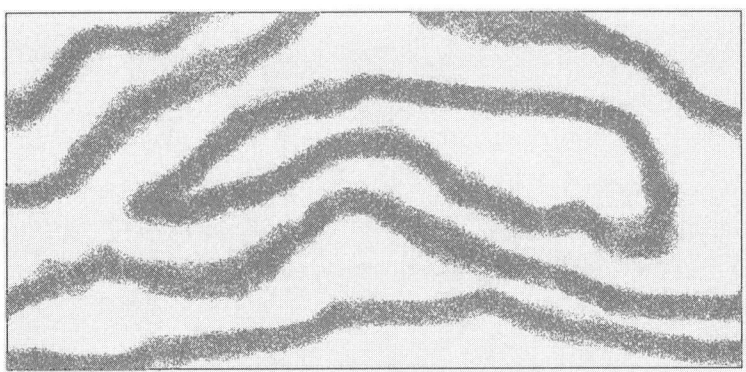

Figure 5.2 Grazing incidence interferometer image of a typical glass slide.

Porous materials are usually vacuum impregnated with the same epoxy resin used for bonding. Vacuum is also used to compact loose materials such as sand or soil prior to impregnation (see Figure 5.3), and similar methods of analysis apply. Diverse materials such as coal, teeth, and bone are impregnated to stabilize samples.

Slides prepared for lapping may also be polished to improve images for analysis. The glass slides provide an ideal mounting plate and groups of up to six samples can be polished simultaneously on a carousel using a two-stage process involving 3- and 1-μm polycrystalline diamond, which minimizes the undercutting of soft portions of the rock structure. The final thickness can be reduced to less than 15 μm (590 microinches) for specialist applications, such as fluid inclusion studies [1].

In geological and oil exploration, cylindrical cores up to 6-inch (150-mm) diameter form a large proportion of the test material. The methods and equipment described above can process samples of this size. Analysis can either be carried out by cutting the core into suitable pieces for thin section preparation, or

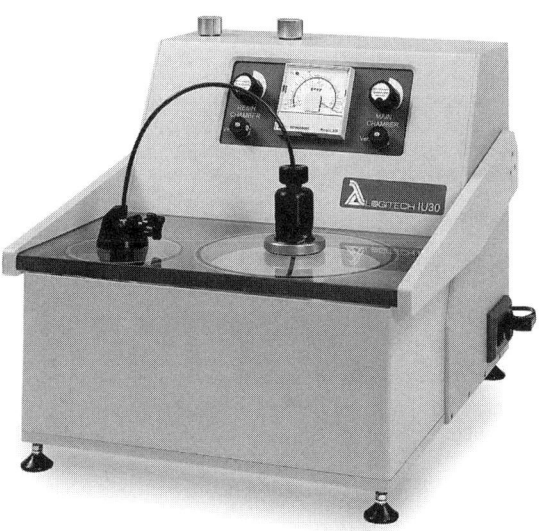

Figure 5.3 Vacuum impregnation unit for porous materials.

by simply polishing the surface of a cross or longitudinal section of the core. Alternative methods such as MRI scanning and electrical conductivity analysis, can now be used, but providing the section of core can be clamped inside a hard-faced ring up to three-fourths the diameter of the available lapping and polishing machine; large areas can then be prepared for microscopic and thin section analysis.

The glass slide shown in Figure 5.2 is typical of many used for mounting rock samples for processing into thin sections. It would be rejected, as it has a hollow up to 4-μm deep. (To prevent both sides of the glass slide producing simultaneous, superimposed interferometer images, the remote side from the reference flat would have a coating of grease to cancel its image.)

The best process is then to select the flatter side by this method, mount the slide on this side by vacuum, and then lap the worse side flat to provide the optimum surface for epoxy mounting of the sample of rock.

Two chambers can be seen under the closing glass flat in Figure 5.3. The RH chamber contains the rock sample which is *outgassed* under vacuum for several hours, even days, before epoxy resin (also outgassed) is allowed to flow under atmospheric pressure from the LH chamber to the right, to fill a small disposable aluminum foil box which surrounds the samples and defines the size of the finished block. Admitting air into the RH chamber then forces the resin into the rock before it hardens. The resulting impregnated block can be sawn to size and processed just like a normal rock sample.

5.2 Hard Materials

This category is reserved for materials harder on the *Mohs* scale than 7, starting with silicon carbide, and moving up through the gemstones, silicon nitride, boron carbide, and finally, diamond.

5.2.1 Lapping Hard Materials

Lapping of hard materials can be difficult, simply because the abrasives available are not much harder than the sample. Silicon carbide, *sapphire,* and ultimately diamond in various forms are required in certain types of devices, both in the structures themselves and as substrates for subsequent processing and *epitaxial* growth.

One property of the hardest abrasives, including diamond, stands out. It enables material to be removed, by lapping and polishing, from the most difficult substances. As crystalline materials, they have different hardness along differing axes of growth; if the grains of abrasive can be made to spin, then the hard axis can usually cut a sample of even the same material. This spinning action is familiar from previous descriptions of the lapping process.

For the hardest materials, preventing the edge of the sample sweeping the abrasive out of the way, and keeping the grit rolling is simply a matter of adjusting the sample load to within very fine limits. The very latest technology makes this much easier and enables lapping to continue under the most challenging of conditions. It is critically important to continue lapping until the finest possible lapped finish is obtained, if the polishing process is to take place in a finite timescale.

As harder samples are processed, it becomes increasingly difficult to change the shape of the surface during polishing, so the shape established during lapping is more important. A brass plate in which the shape is adjustable is very useful in trimming the sample surface shape during fine lapping. The set up is similar to that shown in Figure 2.10, and described in Section 2.1.

In Figure 5.4, the sample mounting face is being lapped into the drive ring to ensure parallelism of the sample. The test block on the left helps to prevent impregnation of the plate with diamond from the drive ring, and assists with maintaining flatness of the plate and mounting face. The jig is capable of the very fine

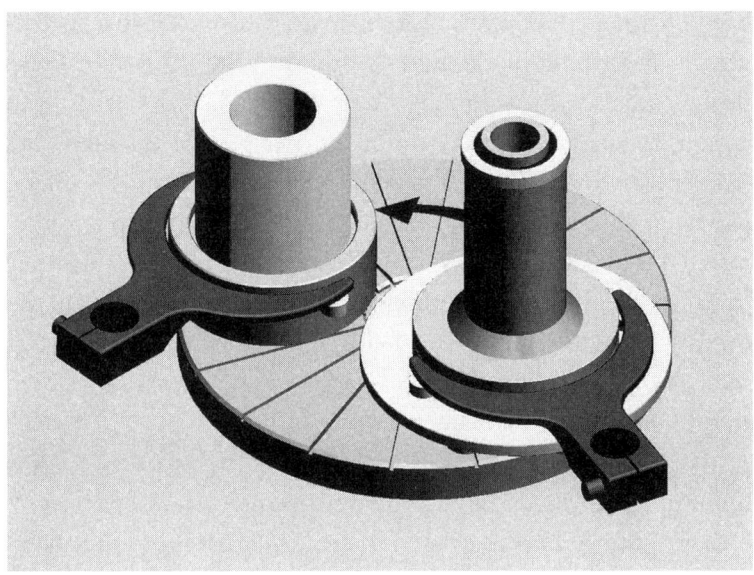

Figure 5.4 Specialist diamond lapping jig (on the right).

load control (down to as little as 10 g/sq cm or 0.142 lb/sq in) which facilitates lapping of hard materials.

5.2.2 Polishing Hard Materials

Finishing of hard materials usually requires diamond abrasives, and here the principles described previously still apply, apart from the fact that the abrasive is available in a range of forms, which provide alternatives for both lapping and polishing.

Natural diamond provides monocrystalline abrasives which are extremely tough and are suitable for lapping. Artificially prepared polycrystalline diamond, on the other hand, is highly stressed and readily breaks down into smaller particles, making it ideal for polishing. (In fact polycrystalline diamond on a paper- or plastic-based polishing pad is perhaps the nearest to a universal polishing combination and can be used in a wide range of situations [2].)

For hard materials however, it is often impossible to prevent local lapping of the sample occurring in hollows or faults in the surface, which prevent a uniformly polished surface from appearing, and resort to chemical slurries such as *alkaline silica sols*. These are significantly alkaline (these sols are only stable at pH range 8 to 10 [2]), and the alkalinity seems to help stock removal. However, long polishing times can be expected, and it may be worth investigating methods of allowing the process to run overnight. Increased load may help, and Figure 5.5 illustrates two methods of achieving this: load added to the jig and a grooved surface on the polyurethane polishing plate.

Care is necessary in this situation not to distort the sample mounting plate, and for many hard materials, SAE416 steel plates bolted to the jig instead of vacuum retained glass mounting plates

Figure 5.5 Polishing jig with extra load weight on a polyurethane plate for hard samples.

are used. Samples may benefit from epoxy instead of wax bonding, as the steel plates and sample materials can often be heated to high enough temperatures to allow breakdown of the epoxy and removal of the sample after processing.

This is the most frustrating aspect of polishing natural, and in particular, *CVD diamond films*. If any part of the sample surface should be at a height above the polishing plate greater than the diameter of the abrasive particles, then any free particles in this space will start to roll and lap the surface by a process called *undercutting*. At the same time, any areas in direct contact with the plate surface will be polishing, a process which we have seen before and is much slower than lapping. With CVD material, there are almost always areas between individual grains, which are prime sites for this to happen. Once again, every sample is different and it is up to the operator to recognize when local lapping is occurring.

Both the grooved plate and the loading weight increase the effective sample load. In a situation where the sample surface is not flat, or has fault lines which could allow a free abrasive particle to roll, this combination, together with an alkaline silica sol solution is one proven way to make progress (the setup will look similar to Figure 5.5). However, polishing times are very long and the sample program must allow for days rather than hours. Stock removal is very load-dependent, and the lands of the grooved plate that are in contact with the sample should be made as narrow as 1 mm and the plate shape significantly convex. Consider also the use of harder plate materials such as soft metal (tin) and FR4-printed circuitboard material [3].

5.3 Water-Soluble Materials

Special fixtures for jigs make it possible to control face angles on most materials; however, as will be seen in Section 5.5, many electro- and acousto-optic materials are water soluble and present special problems. Water in the abrasive suspension fluid acts to cause all the problems associated with chemical polishing, severe edge rounding and high stock removal rates, to mention but two. On the other hand, if fluids containing little or no percentage of moisture can be used then almost all normal process methods apply. For instance, the laboratory grade of *ethylene glycol*, one of the most common processing fluids, often causes problems, while the *Analar* grade, which contains much less water, can be used for most water-soluble samples. Paraffin (kerosene) is often used both to store water-soluble crystals and as a polishing fluid, if local regulations allow its use under lab conditions. Many water-soluble crystals can be recognized by the "ite" suffix in their name (e.g., fluorite and calcite).

However, in addition to water, extreme sensitivity to heat, static electricity, and stress are factors which can make these samples more than difficult to handle. It is also very difficult to image faces of this type of material using interferometers, often because extreme edge rounding results in massively convex surfaces. Even atmospheric humidity can cause this to happen during polishing. Sometimes, the best start can be made by using a soft nap pad lying on a surface table in as dry an atmosphere as possible, and simply rubbing the sample by hand until an interferometer image can be obtained. (Developing an even, figure-eight motion of the sample on the pad is the best way to retain the surface shape.) This image then enables a more accurate setup to be made, together with confirmation that any bonding or mounting method used has not caused distortion of the face. If in doubt, do the absolute minimum to a sample until you can be sure you can test, and therefore control, the important surfaces (Figure 5.6).

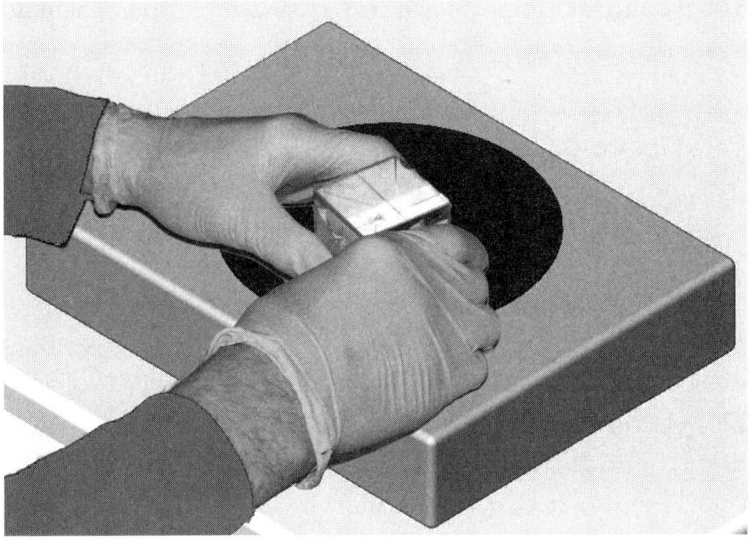

Figure 5.6 Hand-processing a water-soluble sample.

Sawing these materials can be done as usual providing a nonaqueous fluid is used. If this is not possible, then a wire saw (Figure 6.11) using a thread instead of a wire can be very effective, with water used as a minimal lubricant and cutting fluid. If a satisfactory polishing process is found, then lapping after sawing is rarely necessary.

Potassium di-hydrogen phosphate (KDP is the most common water-soluble material) is in demand because its nonlinear characteristics enable it to control light waves passing through it in a similar way that transistors can control electrical current. However, it is extremely soluble in water, which can absorb over 20% of its volume of KDP.

Outside the area of research, the technique of diamond turning and milling is often used for this type of material; the process can be carried out automatically in a completely dry environment, and production quantities justify the costs involved.

Samples such as the one shown are very susceptible to mechanical, thermal, and electrical damage.

5.4 Electro-Optic Materials

5.4.1 Infrared and Electro-Optic Materials

Some (mainly crystalline) materials, such as quartz and calcite, naturally refract light differently for different polarizations (*birefringence*). There are also a number of crystals that are not birefringent naturally, but become so in response to applied voltages. This electro-optic effect gives us the ability to control light in electro-optic devices [4].

Electro-optical materials are split into two classes: amorphous (mainly glasses) and crystalline. For many years, up until the invention of the laser in 1960, glasses prevailed, but as most were opaque at wavelengths above that of natural light. Devices operating above this wavelength in the infrared range were subject to restricted development, until windows and other optical components were made available in materials which were transparent at these wavelengths. Largely crystalline, these materials placed new demands on laboratories attempting to process them. Processing laboratories can expect to be asked to produce both samples and optical elements in any material and must research their characteristics before starting on what may be a very rare, expensive, and possibly toxic material.

A few of the glassy types of infrared transparent materials, called chalcogenides, were developed but were not a commercial success. More successful were two polycrystalline materials made by hot pressing from powders under the trade name Irtran, zinc selenide, and its stronger but less transparent brother, zinc sulfide [5]. Zinc selenide suffers from a problem not dissimilar to diamond in that polishing can fail to progress due to self-propagation of small faults in the surface, and it must be polished using the finest possible abrasives. *Magnesium fluoride, calcium fluoride,* and *cadmium telluride* also followed with limited success and application. Improved versions of these materials are now made by *chemical vapor deposition.*

The most universal infrared material is *germanium.* Although not so strong at high temperatures as zinc selenide, germanium has good transmission (with antireflection coatings) and a very high refractive index at the common IR laser wavelength of 10.6 μm (417 micro-inches), which makes for light and efficient optical systems.

Silicon has emerged as the front runner as substrates for electronic devices, infrared mirrors, and some electro-optic devices. Although it is very common in the earth's crust, it is very widely distributed. Its availability and cost is affected by the need for extreme purity. Many different refining processes are used for specific applications. Its infrared transmission is only useful in the wavelength band from 3 to 5 μm [5].

As an IR transmission and semiconductor material, *gallium arsenide* is now second only to silicon. Its raw materials are readily available as biproducts from aluminum and zinc production. Its use has grown with the demand for diode lasers, although IR optical uses are less common, except for high-power applications where it out-performs zinc selenide.

Materials rejected from this family due to insufficient IR transmission have found favor in recent years because of their good transmission at the other (ultraviolet) end of the visible spectrum. *Lithium fluoride* is one of these, and plays a part in the ultraviolet optics necessary to enable high-resolution lithography to be performed in the recent push to smaller and smaller semiconductor devices [6]. Fortunately, it is not particularly water soluble and fairly hard, so is reasonably simple to polish. Its brother, *magnesium fluoride,* is used in similar applications but is more often found in antireflection coatings for other optical components.

5.4.2 Processing Infared and Electro-Optic Materials

Electro-optic crystalline materials provide a range of challenges for optical processing. Interest in new laser sources has increased with the ability of some crystalline materials to double the frequency of a laser beam, bringing light from laser diodes into the visible range and making possible the utilization of chip lasers for ultrafast optical computing. *Lithium niobate* is one of the most readily available of these, and is hard and comparatively easy to polish with colloidal silica.

In most crystals, each of the prominent crystalline faces belongs to one of three planes that intersect along the crystal axes. When one fabricates an electro-optic device, the crystal faces must be cut at certain specified angles to these axes which are identified by x-ray alignment. Cutting and polishing the crystal then depends on being able to accurately transfer this angle to the polishing machine or jig. Attempts have been made to create angle indexing features small enough for use on normal polishing jigs, but insufficient angular resolution in the end frustrates them. The usual solution (given the fact that the angles required are usually constant for a specific material) is to use accurately machined sample mounting plates incorporating the required angle.

Devices using acousto-optic materials (e.g., lithium niobate) require planks of material which can be subjected to the modulation frequency, having optical fibers coupled to suitable faces. Specific angles and perfect sharp edges are involved here, requiring methods which not only present the correct angle but also reduce edge rounding. Specialist fixtures and procedures make this possible (Figure 5.7).

Shown here is a jig with a stack of lithium niobate planks wax-bonded together for edge polishing. With careful attention to detail it is possible to polish adjacent faces at a specific angle with virtually no edge roll off. In this case, the sample mounting

Figure 5.7 Special fixture for clamping and alignment.

plate under the fixture can be prepared, ground accurately to the correct angle (Section 4.8).

References

[1] www.corex.co.uk/thin_section_analysis.php#.

[2] www.kemet.co.uk/product.asp?productID=1800&prodCat=Polishing.

[3] www.americanepoxy.com/g10fr4sheets.html.

[4] http//.scholar.lib.vt.edu/theses/available/etd-061899-103951/unrestricted/cain1.pdf.

[5] Kaplan, H., "What's New in IR," *Photonics Spectra*, June 1986.

[6] www.crystran.co.uk/lithium-fluoride-lif.htm.

6

Specialized Techniques

6.1 Diamond Machining: Introduction

Until now, the techniques described have been logical, first principal operations that, in practiced hands, can produce commercially relevant results. The temptation is to regard these as easy, and in the last century, literally hundreds of small companies started and failed in the attempt. This was usually because the process of scaling up production to produce high-quality components in sufficient quantities to pay for normal overheads became too difficult to maintain. At the other end of the scale are the research-based operations where the demands are for nanometer-level accuracies and atomic-level finishes. These can originate from the areas of nuclear or astronomy research, which produce sufficient funding to pay for machines that are as sophisticated as they are expensive. In this area, it may only be necessary to produce several individual high-quality components to justify a huge capital expenditure.

6.1.1 Diamond Machining of Ductile Materials

Reflective optical components make up much of the market for diamond machined components. Aluminum and *electroless nickel*-coated aluminum mirrors largely meet this demand, except in the space arena, where the demand for extreme weight reduction leads to processing risks being taken with beryllium, whose dust is toxic. Its diamond machining properties are excellent and are combined with stability, stiffness, and minimal weight.

Many of the simple processing techniques previously described are still relevant to diamond machining, particularly in the area of low-stress sample mounting and alignment. However, the basic machine requirements for removing material with a single point diamond tool, which is the approach mainly adopted, are much more demanding. Apart from research applications, these methods are beyond the reach of all except well-funded companies who wish to enter this market on the back of a high-volume specific application.

An idea of the difficulty may be gained from realizing that this type of tool can only take a cut of a few microns depth without shattering. The knock on effects of this on the machine components demand indexing of cuts to an accuracy better than one-tenth of this. Freedom from backlash is required, and the ability (in the case of a *CNC machine*), to follow a two- or three-dimensional path to the same or better accuracy. This, in turn, creates the need for high orders of stiffness in the components involved with huge effects on the weight of the machine structure, and the need for total balancing of any rotating components. This must be combined with first-class isolation from environmental vibrations, which will otherwise be reflected in the path of the tool. This is simple but costly to achieve by mounting the machine on a cube of concrete, with side the largest plan dimension of the machine. This is often housed in a pit with the machine plus a concrete block on rubber tires fed with compressed air.

The net effect is that a diamond machining installation will have a capital cost at least one order of magnitude higher than that of the machines previously described, and this will be reflected in the cost of any machined components. Both reflective (usually aluminum or electroless nickel) and transmissive (usually plastic) and even very hard (silicon carbide) optical components can be machined. The main issue of machine stiffness is common to them all, but more severe the harder the material. For example, although a single point diamond tool can true up a plastic sample which is running slightly eccentric, on a harder material the eccentricity remains, simply because inadequate stiffness in the machine structure or sample mounting allows the work-piece to move away from the tool as it tries to take an intermittent or heavier cut. The result is that many diamond turning and milling operations are multipass processes where initial truing and roughing cuts are made on standard CNC lathes and mills so the time spent on the expensive machine is minimized.

The basic configuration is similar to standard workshop lathes, mills, and CNC machines but there the similarity ends. To avoid vibration generation and transmission to the work-piece, the machine base is often in granite, on an air suspension mechanism. Slides too may be in granite. Drive motors are often supported on air bearings, and CNC indexing and control have often limited travel because, to reach the high standards of accuracy required, the available resolution is concentrated in a short length. These tolerances are now at previously unattainable levels. For instance, 0.1 μm (4 microinches) for lengths and diameter, shape errors within 50 nm (2 microinches) and surface finishes down to 5 nm (0.2 microinch) Ra [1].

6.1.2 Diamond Machining of Brittle Materials

The concept of single point diamond turning described earlier falls down when applied to glasses, although many electro-optic crystalline materials can be diamond turned. However, the same name is applied to procedures used in the ophthalmic industry for production of spherical and aspheric surfaces in glasses and in ceramics, which are then used as moulds for plastic lenses.

Instead of a single diamond tool with a very accurate nose radius, this industry uses a thin wall tube of steel filled with a graphite material rotating at about 30,000 rpm. This has a layer of diamond particles brazed to the surface, and the rotational axis of the tube is angled to the work so that it is, in effect, single point contact of tool and work.

This is more akin to ultrafine grinding than turning, and linear as well as rotational movements are applied to the tool using CNC control so that aspheric surfaces can be generated. The resulting faces are polished using rotating flexible pads.

Now that plastic lenses are prevalent, methods are changing to single point turning rather than fine grinding, in this case of the moulds, reducing or eliminating the need for smoothing operations. The main issue here is that to amortize the cost of manufacture of mould manufacture, a production run of at least 10,000 identical parts is required [2].

6.2 Sawing

Cutting of fragile crystalline and glassy materials is very different from metal. Whereas metals require a sustained pressure from a sharp edge at least 2% harder than themselves, these materials will surely shatter in this regime and require much more gentle treatment. Simply holding the material for cutting can be difficult but fortunately there are similarities between lapping and cutting which make the job easier to understand. The range of materials to be covered is large, from different hardnesses of rock, (flint to granite to chalk) to the full range of glasses, and then the fragile electro-optic materials, the characteristics of which can be altered by too much stress. This means that very different methods of sawing are required, starting with that for the most fragile materials.

6.2.1 Wire Saws

It is no coincidence that saws and lapping machines are often housed in the same lab. In fact, one method for very sensitive materials involves lapping along the path of the cut with hard abrasive grit slurry on a moving wire, which, in effect, takes the place of the lapping plate. A very sensitive control system is required for the wire, with accurate sample location and path geometry, plus plenty of time to make the cut. Imagine the minute cutting load to achieve 200 g/sq cm.(2.85 lb/sq inches) pressure on a wire only 0.2-mm (0.008-inch) thick. However, this is one of the most secure methods for this type of material. As shown in Figure 6.1, the sample is usually held by wax or epoxy resin onto a bar of soft ceramic material into which the wire can cut at the end of the operation. This allows the individual wafers to be broken off without damage.

Figure 6.1 shows a small lab saw. Similar principles are applied to much larger units and may use wires with diamond grit brazed to the surface, rather than loose abrasive lapping. Larger units still cut rock and concrete, often with steel re-enforcement in the construction and mining industries, one of the most impressive being the cutting up of collapsed flyover road sections after an earthquake caused collapse. Here the cutting grit is contained in large pellets of *bronze* or steel brazed to multistrand wire rope.

The left knob on the panel controls the wire speed, while the right one allows control of the very small load the sample exerts upwards on the wire. Slurry drips onto the wire from the drum at the rear. The 75-mm (3-inch) diameter *boule* of material in the center is the sample, the material in this case being lithium

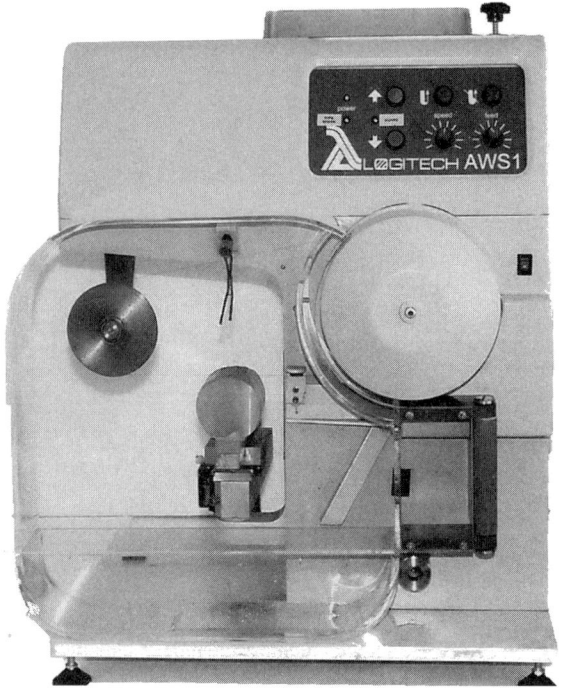

Figure 6.1 A wire saw for cutting fragile materials [3].

niobate. Although it takes a long time for the wire to lap through, very little stress is applied to the material as is appropriate for electro-optic applications (Section 5.4).

6.2.2 High-Speed Saws

To speed up this process for less fragile materials, the grit is held fixed, often in a bronze matrix, and rotated past the sample in a process more akin to grinding than to lapping. The relative speed between grit and sample is in the region of 5,000 ft/minute, instead of 100. (This is the principle used by most saws for hard materials.) Like grinding however, it produces high levels of subsurface damage in the sample, which usually requires lapping before polishing.

Specialist saws for geology, like the one shown in Figure 6.2, are able to locate and thin 150 × 100 mm (6 × 4 inch) rock slides to the point where they can be lapped down to 27-μm (1-mil) thick and viewed under a microscope in transmitted light (Section 5.1).

This type of saw is eminently visible in the building industry where its huge advantage is in its absence of sharp teeth, which gives fewer safety problems. When this roughing saw is provided with an *accurate* blade and *precise* control system, an excellent and very capable saw results. As in diamond conditioners (Section 2.6), the blade can be sharpened or dressed using an abrasive. In this case, it is in the form of a rectangular stick into which repeated short cuts can be made. These saws are provided with a liquid flow for cooling and lubrication, but as the speeds are very high, only a small quantity of fluid is required, otherwise hydrodynamic forces can literally stop the blade cutting. Fluids other

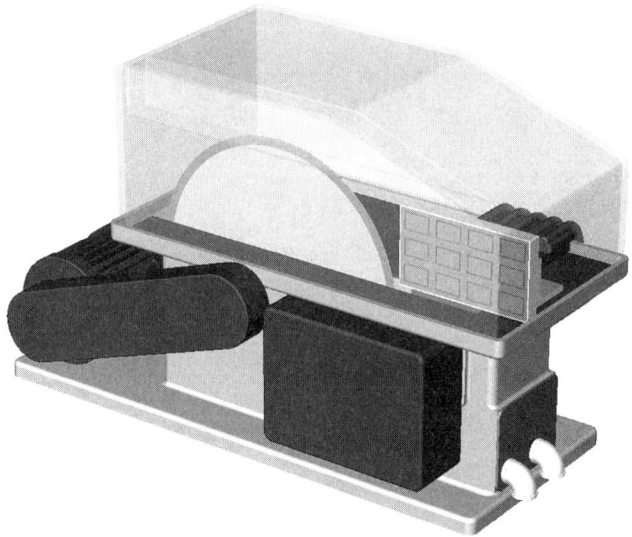

Figure 6.2 High-speed cutoff saw with special features for geological samples. The blade is 300-mm (12-inch) diameter.

than water, either a cutting oil or *ethylene glycol,* are used for soft and water-soluble materials like marble and chalk.

As shown in Figures 6.2 and 6.3, there are several configurations of saw with this general format. The unit in Figure 6.3 has two-axis *CNC* control, and the stainless steel pillar under the sample provides vertical and rotational indexing. In addition, dedicated dicing saws have small stiff blades for separating semiconductor devices on wafer dies and advanced CNC control to minimize losses of valuable components. These highly automated versions of this type of saw have blade diameters as small as 50 mm (2 inches) and spindle speeds as high as 50,000 rpm. They are programmable with typical step accuracies of +/− 5 μm (200 micro-inches) and in the blades, the width of cut can be as low as 20 μm to minimize material loss between dies on the wafer [4].

6.2.3 Annular Saws

Annular saws literally stretch a steel diaphragm as thin as 40 μm (0.015 inch) with a hole in the center, to the point where the hole

Figure 6.3 A high-speed saw for dicing semiconductor wafers. The sample being cut is a 100-mm (4-inch) diameter.

enlarges sufficiently in diameter, in a perfectly round shape, to support the internal rim of abrasive. This can part extremely thin slices from a boule of semiconductor material. The inside edge of the hole is coated with a diamond film bonded with bronze or nickel alloy.

This is the preferred method in industry for tough materials, like silicon, which can withstand automated or semiautomated processing. The unit shown has sophisticated controls enabling sequential cuts to be made in a boule of artificially grown

Figure 6.4 Annular saw with a 100-mm (4-inch) diameter internal hole.

crystalline material. At the end of the cycle, the stack of cut wafers is removed complete from inside the drum and the individual wafers broken off the ceramic bar.

The hole in this blade, shown in Figure 6.4, is approximately 10 cm (4 inches) in diameter, and dictates that the outside diameter of the drum is at least 30 cm (12 inches). Much larger units, the size of a room, can handle boules of material this diameter or larger.

The hexagon key bolts seen round the periphery of the drum are used both to grip the steel diaphragm and to tension the blade. The tension is such that the center hole can increase in diameter during setup by 100 μm (0.004 inch) and be made perfectly circular to within 10 μm (0.0004 inch).

References

[1] www.precitech.com/product-overview/nanoform-250-ultra/.

[2] Jailie, M., *Opthalmic Lenses and Dispensing*, 2nd ed., Garfos S.A., University of Ulster, Spain, for Butterworth-Heinemann, 2003.

[3] Logitech WS10, www.logitech.uk.com/ws10.asp.

[4] Karow, H. H., *Fabrication Methods for Precision Optics*, New York: John Wiley & Sons, 2004, p. 367.

7

Surface Finish

7.1 The Lapped Surface Finish

The surface finish obtained by lapping depends on the abrasive grain size and the hardness of both plate and sample. However, in general, a finish from an alumina grit, nominally 9 μm (354 micro-inch) in diameter or 600-grit silicon carbide, will give a surface finish around 300-nm Ra (120 microinch Ra). This is the roughest finish that can produce a contour image in a grazing incidence interferometer of 4-μm nominal wavelength and is a suitable starting point for most polishing processes, including those used in geological thin section preparation. Coarser 50-μm (2 mil) grit will produce a finish up to 1 μm (39 microinch) Ra (Figure 7.1). The finest lapped finishes, down to 50 nm (2 microinch), are obtained using 3-μm (12 microinch) monocrystalline diamond on a brass plate with annular grooves.

Figure 7.1 A lapped gray surface, under the microscope 200× (0.5-mm or 0.02-inch diagonal).

7.2 Subsurface Damage

The depth of damage caused to a sample by lapping determines how much material must subsequently be polished off to give a damage-free surface. Physical measurement of the depth of damage has been attempted, but without very expensive x-ray diffraction equipment, or electron microscope examination of cross-sections, this has to date proved impractical, and lab staff have to rely on a rule of thumb.

Looking at the stylus trace of the lapped finish in Figure 7.2, the Ra (average) finish is about 1.5 μm (1,496A), but the peak-to-peak depth of the finish is almost 10 times this (14,026A) . So to remove the finish, you will have to remove at least 15 μm (0.6 mil). To allow for further subsurface cracking, the rule of thumb would be 3 times this again [i.e., almost 50 μm (0.002 inch)] or 30 times the Ra finish. As an even rougher guide, this is perhaps three times the abrasive grit size used in lapping. A finish produced by grinding, unless subsequently lapped, requires

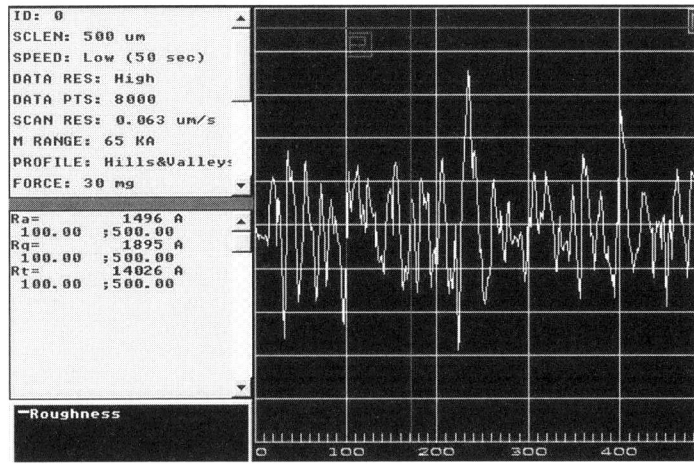

Figure 7.2 Trace from a stylus instrument of a similar surface to Figure 7.1.

twice this amount to be removed during polishing. Techniques to remove this damage while maintaining surface form are described in Section 3.4.

7.3 Understanding Surface Finish

Understanding surface finish measurement begins with visualizing the shape of the whole surface: flat, spherical, cylindrical, conical, or a combination. The fact that every part of the surface, however small down to atomic size, can be viewed in the same way leads to a surface finish that is random unless patterned by design. Most workshop finishing methods, such as milling, shaping, and grinding, disturb this random surface by imposing a lay to the finish which follows a geometric shape generated by the machine function (straight, curved, or circular). Lapping and polishing should generate random finishes. This is why sweep movement on a polishing machine is desirably independent of the main plate speed. Measurement of the surface finish occurs normally in a straight line at right angles to any lay. It is measured over a specific distance called the sampling length.

7.3.1 Cutoff

Machines using a stylus for measuring the surface are able to discriminate between waviness and roughness by imposing a limit on the spacing distance of peaks along the measurement line. Peaks and valleys along this distance comprise roughness, whereas a line joining peaks at the ends of the sampling length and continuing over further multiples of this length traces the waviness. The spacing distance limit is called "cutoff" and is a machine-specific distance. This quantity must be specified before making a measurement but is less relevant on random surfaces than ones with a specific lay. One specific machine type uses the surface on test as a reference by pivoting the arm carrying the stylus on a shoe which rests on the surface. The length of the shoe can be considered as the cutoff length. In software terms the cutoff length is the distance below which surface finish features are included in the surface finish computation. Above this length, any features are ignored and designated part of waviness rather than roughness.

7.3.2 Stylus Radius

Measurement is by an industry-standard stylus whose tip radius radically influences the result. In Figure 7.1, a diamond stylus with a tip radius of 1 μm is preloaded against the surface with a load of only 30 grams and drawn across the surface for a distance of 500 μm (0.02 inch). The movement of the stylus is amplified and recorded on the vertical scale, which runs (in Figure 7.1) from + 1 μm to -1 μm. More recent atomic force microscopes (AFMs) use a stylus radius down to under 10 nm, produced by vacuum techniques often employed in semiconductor device manufacture. These instruments, however, have a limited sampling length in the order of 150 μm. They can produce three-dimensional scans of this size, but can be time consuming to set up and use. If their accuracy was to be maintained over the full instrument travel, say, 50 mm (2 inches) (or 100 times the actual scan length), then the mechanism controlling the movement of the stylus (the reference component) would have to be accurate to one-tenth of the vertical resolution of the instrument. (A generally accepted norm is that a measuring instrument should be ten times more accurate than the resolution of the measurement it makes.) This means that the slide or other mechanism guiding the stylus movement has to be absolutely flat to within the size of 1 atom. To put this in perspective, the best optical reference flats are generally flat to 250 atoms (one-twentieth of the wavelength of light). So for this reason, in general, the more sensitive the instrument, the more restricted the travel. This is also related to the computer hardware, as many of the systems used in such machines are limited to 16-bit data (16-bit represents 2 to the power 16 or 65,536 unique data values), so the smaller the data unit (Angstrom, micro inch, micron, and so forth), the less length can be covered. A micron accuracy instrument could then have a maximum travel of 65.5 millimeters. Because of this, such instruments are rarely used to examine surface shape as opposed to surface finish. Although stylus profilers with much larger travel, up to 300 mm, are available, they have correspondingly less *resolution* and *accuracy*.

Figure 7.3 shows the surface trace from Figure 7.2 with the horizontal and vertical scales approximately equalized. Superimposed on the right is the radius of a typical diamond stylus (about 1 μm) used to make this trace by travelling approximately the length of the horizontal axis (0.5 mm or 500 μm). In contrast, the probe of an AFM is shown on the left, having a tip radius of about 10 nm. Typical travel of the AFM probe would be 150 μm. The function of these two probes is different (the stylus presses on the surface with a very small finite load, approximately 50g, whereas the AFM tip is balanced just above the surface by atomic

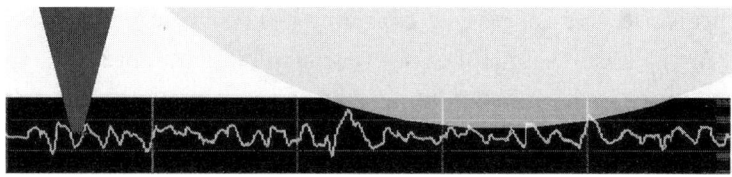

Figure 7.3 Trace from Figure 7.2 with horizontal and vertical scales equal [the horizontal axis is 500-μm (0.02-inch) long].

repulsion forces). It can be seen that the surface finish trace from the probes of these two instruments will be very different.

In order to define this in a single figure for comparison purposes, the average and measured finish can be represented in several ways. Common to them all is the centerline which is the line above and below which the area between the line and the curve is equal on each side as shown in Figure 7.4. Software can be used to neglect and correct any out of straightness of the centerline during a scan. (In effect this amounts to ignoring any surface shape variation which has a periodic length greater than the cutoff value, say, 0.1 mm.)

The roughness average (*Ra*) is the most commonly used value and is a measure of roughness based on the magnitude of this area. In mathematical terms, it is the mean of the sum of all positive (or alternatively negative) values of the vertical axis from the centerline, over the sampling length being considered. It then has a single value with a length dimension (usually microns in heavy industry, nanometers or microinches in semiconductor manufacture, and angstroms in research). In Europe (ISO standards) it is often called centre line average or CLA.

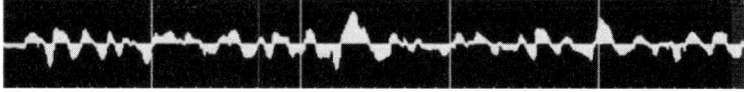

Figure 7.4 Trace from Figure 7.2 with areas equalized above and below the centerline.

Typical industrial workshop values of Ra are from 0.0125 μm (ISO designation N5) to 25 μm (ISO designation N11) representing from polished to sawed surfaces (from 0.5 to 1,000 micro-inches). The coarsest finish that can normally be imaged in a grazing incidence interferometer is about 0.4 μm (16 micro-inch) Ra, which represents the finer end of the normal range of turned, ground, lapped, or honed finishes.

In semiconductor work, Ra values range from about 0.001 μm to 1 μm, from highly polished to coarse lapped finishes (from 0.04 to 40 micro-inches). Note that 0.001 μm is 1 nm or 10 angstrom units. Therefore, the finish shown in the Figure 7.2 trace is shown as 0.15 μm, 150 nm, or 1496A, over the one-half-mm scan length.

The alternative value that is commonly met is Rq, which is the root mean square value of the curve. In a curve like Figure 7.2, it can be approximated by the following formula [1]:

$$Rq = \text{Curve Amplitude}/1.414 \qquad (6.1)$$

The amplitude of the curve shown (positive or negative from the center line) can be estimated as approximately 2.8 KA.

$$\text{So } Rq = 2.8/1.4 = 1.9 \qquad (6.2)$$

This compares favorably with the detail computed value in Figure 7.2 of 1,895 KA.

The maximum finish parameter (*Rt*) is simply the maximum amplitude peak-to-valley dimension in the sampling length, and the mean (*Rtm*) is that value averaged over five readings found in the sample length.

Very similar is *Rz*, which is the mean peak-to-valley height where five successive sample lengths are assessed for peak-to-valley height and the sum of these divided by 5.

Reference

[1] www.zygo.com/library/papers/SurfText.pdf.

8

Optics

Up to now, we have been looking at a small sample processing laboratory, such as is found in many university departments throughout the world. Similar labs exist in a large number of quality control departments in industry. These concentrate on producing flat and parallel samples in new or rare materials; or alternatively, in producing quantities of a specific type of sample for test purposes. Taking the step from this to a production environment is a daunting prospect.

Processes for glass optics using lapping, polishing, and grinding are similar to those described earlier with a few exceptions:

- Machines are larger to match higher volumes and workpiece size.
- The higher volumes lead to much faster process times.
- Polishing is often carried out on pitch plates using cerium oxide abrasive.

- Smoothing operations (Section 3.4) are often used instead of lapping, especially on spherical surfaces, where diamond grinding normally generates the spherical face.
- Investment in test equipment can be high because of high volumes.

8.1 Glass

Supply of stock glass for the optics industry tends to be in the hands of a few major firms. Glass is formed when a mix of silica sand (with some additions), is melted and allowed to cool rapidly. It forms an amorphous solid which has not had time to crystallize.

Various additions and impurities in the mix can alter the way in which light passes through a glass or other transparent material. First, the speed of light is altered, slowed down by an amount defined by the material's refractive index. Second, the extent to which different wavelengths of light are slowed down by different amounts is defined by a quantity called *dispersion*. The additions vary with manufacturers and result in a wide range of glasses with differing properties. After many years of confusion due to different codes adopted by manufacturers in different countries, a universal six-figure code now specifies these parameters. For example, 517 642 identifies BK7, a moderately hard and stable optical glass whose refractive index is 1.517 and dispersion index is 64.17.

Lens designers tend to use a standard range of glasses. These are often custom supplied in suitable blanks for finishing and then sawn, cleaved, and ground to size prior to spherical surface generation and polishing, as shown later in this chapter.

As well as working with glasses, optical workers can expect to be presented with many of the range of electro-optic materials described in Section 5.5. This is because most of them are transparent in parts of the infrared or ultraviolet range (shown in Table 8.1), and can be useful optical materials for these wavelengths.

Table 8.1
Electromagnetic Spectrum

Wavelength	Frequency	Feature
1 km	1 MHz	VHF radio
1 m	1 GHz	UHF radio
1 cm		SHF radio
1 mm	1 THz	Infrared light
1 μm		Visible Light
		Ultraviolet light
1 nm		X-rays
		Gamma rays
		Cosmic rays
Wavelengths of the Visible Light Range Expanded		
	380–450 nm	Violet
	450–475 nm	Blue
	476–495 nm	Cyan
	495–570 nm	Green
	570–590 nm	Yellow
	590–620 nm	Orange
	620–750 nm	Red

(Wavelength = V/Frequency, where V = velocity of light)
Source: [5].

However, some of them also have awkward physical properties, such as water solubility and sensitivity to electrical or mechanical damage, and it is the responsibility of the optical shop to research and find ways around these issues.

8.2 Processing with Pitch

The most noticeable machine in any optics shop is usually the pitch polisher. It often occupies a pride of place—and is nearly always turning. Up to 8 ft in diameter, the configuration (Figure 8.1) is not dissimilar to the 1-ft machine shown in Figure 3.1, with three workstations and a conditioning station at the rear. However, jigs are rarely used and are replaced by work rings, slightly larger in diameter than the width of the polishing track. The large conditioning block at the rear is up to 1.5 times the track width in diameter and may be in granite with the face in contact with pitch, covered in a random pattern of round glass plates. Individual work-pieces, or blocks of parts, run in the work rings. Their loading is controlled by the weight of the block on which they are mounted, or occasionally by pneumatic rams mounted above the rings. Where heavy loading is used, many machines have water cooling the baseplate, so that the pitch temperature remains constant. Baseplates are massive, often up to one-fourth the diameter in thickness, and sometimes supported on a film of air to isolate the pitch surface from external vibration. The pitch surface is cast in an annulus up to 50-mm (2-inch) thick after the baseplate is heated, and then the top surface turned flat after cooling to room temperature, often by an integral lead screw on the machine, which gives the whole setup the appearance of a huge horizontal lathe.

A groove pattern is then machined, very often to the operator's choice. Machines with a lead screw have an advantage here; turning flat a continuous scroll, often at about one-fourth-inch pitch, can provide the basis of the groove pattern. To this is added, by hand, a series of radial grooves to aid in the removal of debris. Typically, these are 3/16-inch wide at the surface with a 90° V form, unless small samples are involved where the grooves may be less wide. Once started, the machine can run for days. If stopped, the conditioner is raised to prevent it from sinking into the surface. The start-up time after stopping can be at least an hour to stabilize conditions which affect the sample shape. These include the pitch surface shape, atmospheric temperature, and humidity, as well as weight and position of the sample mounting blocks which are fixed to the back of the work with double-sided tape. Sweep as described earlier is rarely used, the samples running inside the work rings shown in Figure 8.1, and naturally, precessing across the polishing track as the rings rotate. The pitch surface shape is controlled by the conditioner position along the plate radius. It depends on the pitch being able to flow to take up a slightly different shape, and the groove pattern cut into the surface must allow this to happen. In fact, on occasions, the groove pattern is intentionally altered to aid correction of the shape. Up-to-date machines have variable speed drives and in process control of conditioner position, so that test pieces, which may run continuously as well, can be controlled to remain within flatness specification. Slurry is mixed and delivered continuously, and for some work, runs at a finite depth above the pitch, retained by rings outside and inside the pitch annulus.

Why pitch? The pitch/cerium oxide combination is the one exception to the edge-rounding issue which was so much the subject of previous chapters. Pitch for polishing is constantly on the move and conforms to the sample surface as if the two were being sucked together, the conditioner surface constantly moderating the shape. Polishing with it needs simple, old-fashioned skill and

Figure 8.1   Pitch polisher configuration.

frequent attention. However, the absence of edge rounding makes this method so powerful that it is worthy of consideration for any new project. Such a machine can deliver high-tolerance flat or spherical components in large numbers on a continuous basis with zero edge rounding. However, it is very unlikely that any company with less than 12 employees would be able to maintain and make use of such a machine, far less keep it productive with a constant supply of suitable components.

The pitch itself is an equally complex subject. Wood pitch, made from dry distillation (heating) of mostly pine wood, is available in a range of melting points, melting point ranges, and hardness. Then its properties can be further modified by mixing with lighter agents, varying from castor oil to paraffin wax, to serve a wide range of applications, from temporarily attaching lenses to sticks for hand-polishing to coating the inside of a spherical block polisher.

Softer pitches used for polishers are able to flow at room temperatures and constantly conform to the shape of the workpiece, whereas those for blocking and holding components must be hard and stable until heated.

Work-pieces sit in the rings in contact with the inner diameter, which maintains both their rotation and precession across the surface of the polishing track. The outside diameter of the work is protected with tape, and the load controlled by both the suck-down effect onto the pitch and mounting blocks, often in aluminum. These are positioned on the backs of the samples to control sample parallelism, taper, and shape.

8.3 Pitch Alternatives

A wide variety of materials have been used over the years as an alternative to pitch. Starting with wool felt and progressing to up-to-date plastics materials, none has been as successful as closed pore polyurethane foam. Available in sheets of 1.3 to 1.5 mm (51 to 59 mil) thickness and several hardness grades, the combination of polyurethane with either cerium oxide or alumina comes close to the universal polishing method. However, because it lacks the ability of pitch to flow and adapt to the sample surface shape, it always suffers from the usual edge rounding. For spherical lenses, which will be edged and chamfered after polishing, this can often be ignored, but for high-accuracy flats, there is no real substitute for pitch.

Polyurethane is one of the most useful of the *thermoplastic* materials. Others, such as nylon, have been used as linings for spherical polishers formed on the inside of convex tools with aluminum or cast iron male formers in the same way as pitch, and then turned to the accurate radius of curvature, leaving a slightly grooved finish to retain the abrasive slurry. They are capable of higher output and are more durable than pitch on glass lenses, but with a lower standard of accuracy. This is simply because, in general, plastic materials have a coefficient of thermal expansion, some 30 times that of metals, and do not maintain accurate shape if allowed to heat up in the polishing process.

In chemical polishing of wafers, where edge rounding is accepted and allowed for (Section 4.8), a very large range of specialist plastic materials have been developed [1, 2]. These usually have multilayer characteristics to reduce the expansion problems.

Most have been superseded by modern three-layer poromeric pads, which are aimed at volume production. The bottom, strong, fibrous layer in contact with the plate usually carries a pressure-sensitive adhesive with a protective paper transit layer. Next is an increasingly porous polyurethane area which carries the top polishing surface, or nap, building to a total thickness of about 3 mm. The whole has the feel of soft suede leather. In fact the original material was developed for shoe manufacture. It is perhaps sufficient to add that, for optics production, this top layer has to advantage being impregnated with pitch [3].

8.4 Spherical Surfaces

Up to now, this book has dealt with flat surfaces. Graduating to spheres not only opens up the mass market for optical components, but in doing so, makes the company large enough to be eligible for processing a wide variety of every type of sample and material. The difference is in the quantity of tooling, test equipment, and the amount of skill required.

Creating a spherical surface is comparatively easy. The setup is almost the same as conditioning a flat plate, shown in Section 3.2. If a conditioning block is run towards the outside of a flat plate for long enough, you will create a convex toroidal surface. Running it towards the inside creates a concave surface. The condition for the surface to be spherical rather than a toroid is simply for the inner contacting edge of the block to pass through the center of the plate. Making this happen more quickly demands both a grinding operation and a grinding cup ring with a radius nose, so that there is a line contact between the plate and tool (Figures 8.2 and 8.3). This line must pass through the center of the work-piece.

The radius nose of the cup wheel is coated in diamond in a bronze matrix. The lower glass blank rotates slowly on its axis and the cup wheel fast, to give the normal diamond cutting speed of around 5,000 ft/min. The same comments apply to quantity of lubricant as in Section 6.2.2. Excess lubricant can reduce cutting ability.

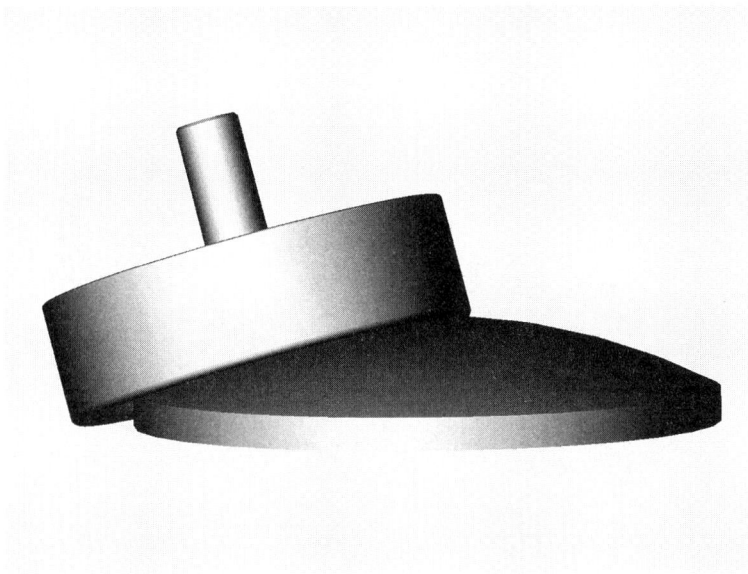

Figure 8.2 Convex spherical generation.

8.5 Blocking Spherical Components

Having to generate each lens of a large batch independently can be very time consuming. (Every piece has to be centered on the slow rotating axis unless held in a custom-made collet.) Identical parts, however, can be blocked into large holders with custom machined recesses, using pitch or wax as the adhesive, and then generated in one operation (Figure 8.4).

Regardless of which method is used, this draws attention to the large quantity of special tooling required for generating and finishing batches of lenses. The advantage, however, is that once blocked, a batch of lenses can remain in the block for polishing, either with a conversely shaped pitch polisher, or with a similar former carrying a polyurethane pad that has been shaped to allow it to conform to the spherical surface. It may also be necessary to create conforming diamond pellet tools for one or more

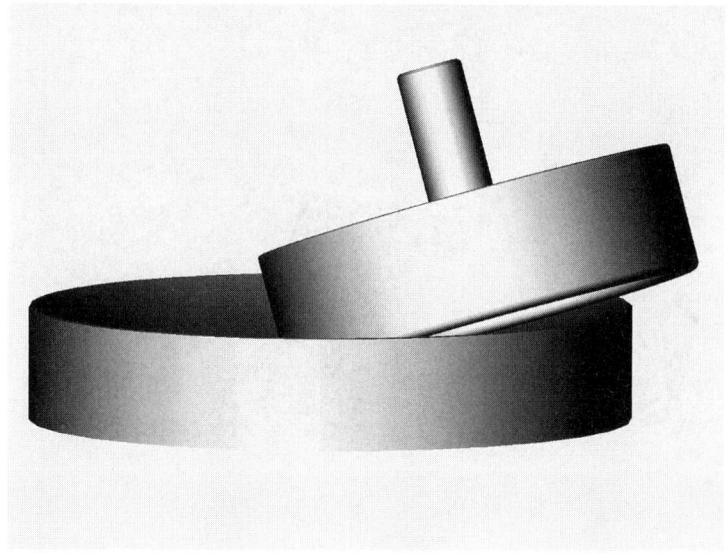

Figure 8.3 Concave spherical generation.

Figure 8.4 Generating a block of convex lenses.

smoothing operations. The total tooling bill, either in purchased formers or in machine shop time to create them, can dominate the process.

Smoothing and polishing are carried out on similar setups; a machine is required which oscillates the smoothing or polishing former over the spherical block. Simpler machines oscillate it in a horizontal plane. For smaller radii, more complex machines oscillate the former around the circumference while the block rotates. Frequently, abrasive slurry is applied with a simple brush or spray bottle, but automated systems are available.

8.6 Specifying Diamond Tooling

The specification of a diamond grinding tool, saw blade, or conditioning block insert has a wide range of conventions, as do the grits themselves. However, bond material and diamond concentration are the main parameters.

Bond material varies from hard (sintered steel) to moderate (sintered bronze) to soft (resin bond). A fourth class of wheel exists where plated grains are held in place as a single layer by an electroplated (usually nickel) matrix. The hard bonds are used with high diamond concentrations (over 20% by volume) for hard work. Moderate to soft bonds tend to be used with lower concentrations (under 20% by volume) for lighter work and finer finishes. Too high a concentration can lead to weakening of the bond matrix, reducing the ability of the part to sustain high stock removal, a sensible maximum diamond concentration being 30% by volume.

The harder bonds need less lubricant, too much leading to reduced output. Softer bond tools tend to use more lubricant to realize better finishes. They are generally used with finer grit and on fragile samples. Almost all tools of this type use monocrystalline natural diamond grit [4].

8.7 Testing of Optical Components

The next most obvious piece of equipment in an optical shop is likely to be the interferometer, often settled on a large air-supported table to limit external vibrations. Most lens types, and virtually all flat components, can be tested on this, and it is likely to be one of the largest capital items on the itinerary. The cost is boosted by special adaptors and lenses (spheres) to suit specific applications, and by the requirement at the top end of the market for air conditioning and filtration in the room. The most common test configurations are shown in Figures 8.5 through 8.7.

In fact, most tests for spherical surfaces involve test plates, the spherical equivalent of a reference flat. The simple test is to bring together test plate and sample, and view the resulting fringes under ambient light. This means that expensive, high-tolerance test plates will have to be stocked for each radius of curvature. Larger firms will stock larger ranges of test plates, to the point where optical designers tend to patronize companies who have suitable test plates with radii suitable for their designs.

This setup (Figure 8.5) is used to measure distortion of a plane wave transmitted through the element under test. It is typically used for windows, filters, and prisms. In addition, glass and other transparent raw material may be examined for homogeneity.

A transmission (Figure 8.6) sphere transforms the mainframe output beam into a precise spherical wavefront for the evaluation of spherical surfaces and lenses. A concave spherical surface is examined for surface figure and irregularity (i.e., the deviation from the best-fitting sphere), by placing its center of curvature coincident with the focus of the transmission sphere.

Convex spherical surfaces (Figure 8.7) are examined for surface figure and irregularity using the setup shown. Both 4-inch and 6-inch diameter transmission spheres are available for these applications.

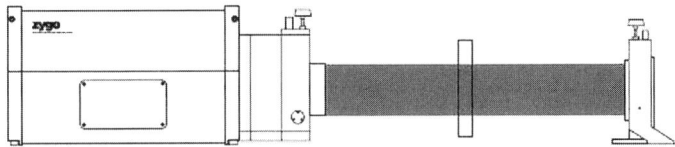

Figure 8.5 Testing of spherical surfaces 1. (Courtesy of Zygo Corp.)

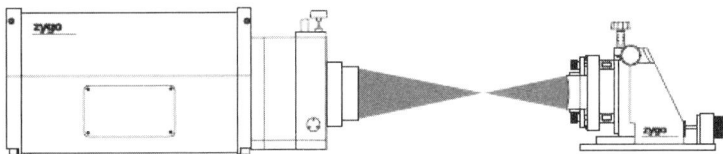

Figure 8.6 Testing of spherical surfaces 2.

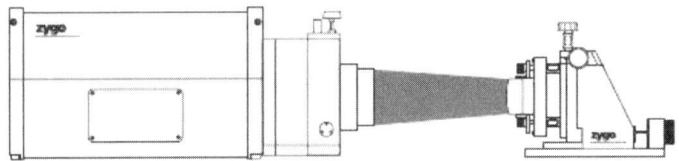

Figure 8.7 Testing of spherical surfaces 3.

The purpose of Table 8.1 is to enable readers to orientate everyday features (radio, colors, x-rays) within the natural range of wavelengths and frequencies. It is not scientifically explicit.

References

[1] http://scholar.lib.vt.edu/theses/available/etd-061899-103951/unrestricted/cain1.pdf.

[2] www.dow.com/products/product_lister.page?industry=1000004&application=1011343.

[3] Karow, H. H., *Fabrication Methods for Precision Optics*, New York: John Wiley & Sons, 2004, p. 223.

[4] Karow, H. H., *Fabrication Methods for Precision Optics*, New York: John Wiley & Sons, 2004, p. 475.

[5] www.tutorvista.com.

9

Semiconductor Device Deconstruction

Sequentially removing layers from a semiconductor device is often necessary, both for quality control and for production engineering purposes. A good quality polishing jig has most facilities necessary for this job; however, this is one of the most taxing of tasks and the methods described in previous chapters provide only the basics for operating at the level of accuracy required. The hints below can be used to reach improved performance in this and many other applications where very thin (submicron) layers are involved.

Consideration of typical device geometry shows why the thin layers cause problems. A 10-mm-square device with a layer thickness of 50 nanometers needs to remain flat and at the correct angle to the polishing plate within 1 second of arc, simply to ensure that the full area of a specific layer can be revealed. This is a tall order, stretching all the procedures and techniques previously described in this book to the limit. Add to this the fact that consecutive layers can be in radically different materials in terms of hardness and electrochemical sensitivity, and you have a very demanding task on hand. The stock removal gauge (Figure 3.1) on a standard jig (even electronic types with 1-μm resolution) is not sufficiently accurate or repeatable to allow location of specific device layers (or even groups of layers) in the direction at right angles to the surface.

It is possible to do the job by hand, simply bonding the device onto a glass or ceramic holder and using one finger to apply pressure between the plate and the device. In the most expert of operators, this may result in revealing a few percentages of the area of a specific device layer, which is visible under the microscope after many attempts. However, careful work with an up-to-date polishing jig can reveal up to 90% of the area of a specific layer with a much higher success rate. To achieve this, methods and procedures rather different from those described so far may be adopted. Above all, the concept of frequently checking the sample under the microscope or loupe (approximately 20× magnification will allow viewing of the whole chip and assessment of the flatness and angle alignment of the polished face) has to be borrowed from the hand processors (Figure 9.1). Effective setup of the microscope is essential to avoid alteration of the color of image features (Section 11.3). This is further aided by using a simple optical flat and monochromatic light (Figure 4.6) to make any corrections.

The first challenge is that of edge rounding. On complete device wafers containing many hundreds of dies, rounding of the edge during polishing is often accepted. On a single device, which may be anything from 1-mm to 10-mm square with tracks and vias within a few microns of the sawn edge, edge rounding will rapidly destroy any attempt to analyze a faulty area under the microscope.

To avoid this, plate hardness has to increase, and with it the likelihood of producing a scratch, especially as the diced edge has, by definition, been sawn. One of the best materials turns out to be that used for printed circuit boards. The composite structure

Figure 9.1 Microscope photograph of partially delayered device at approximately 20× magnification. The sample is approximately 6-mm square. You can see areas where the next silver-colored layer is starting to show.

is suitably hard and the reinforcement, once the surface has been lapped flat, provides the essential "mobility" of the polishing surface. A slightly softer alternative is one of the many grades of thermoplastic-filled fabric boards [1, 2]. The surface is preferably prepared with annular grooves (Figure 5.5), as this reduces the likelihood of scratching (unless the sample is smaller than the available groove width).

The task of identifying where the polished surface lies in the three-dimensional map of the device is down to very careful control of the sample height, together with prior knowledge of the device structure. In fact these parameters, combined with very accurate conditioning of the plate surface (made easier with a hard material) are the key to success. This enables at least the group of layers to be identified, and the microscope then allows identification of progress through the layers in the group. A group is usually bonded by a hard *silicon nitride* or *silicon dioxide* layer and, providing a soft abrasive is chosen (as is necessary to control progress through metallic layers), this will temporarily slow stock removal. Cerium oxide or diamond abrasive is then used to break progressively through the hard layer, hopefully with minimum damage to the metal layer underneath (which may be the specific layer you are interested in).

Three additional features on a specialist polishing jig can facilitate this whole task (Figure 9.2):

1. Angular adjustment;
2. Sample shape adjustment;
3. Micron indexing.

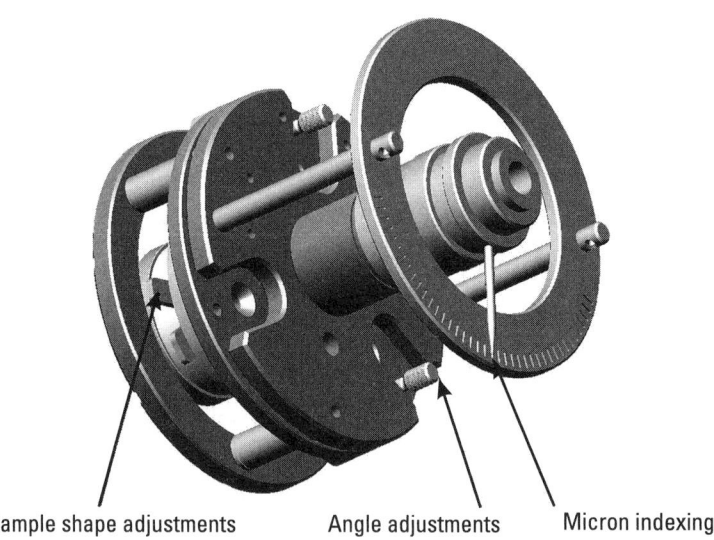

Sample shape adjustments   Angle adjustments   Micron indexing

Figure 9.2 Jig features for semiconductor delayering work.

The angle adjusting plate allows the angle of the sample to be controlled. An adjustable sample mount facilitates adjusting the flatness of the sample without significantly altering its angle. The micron indexing system allows the sample to be lowered or raised in 1-μm steps or less.

The sample shape adjustment can be combined with a simple bonding technique, which avoids stressing the thin device. Instead of attempting to apply load to the top of the part while the wax is molten, which almost always distorts its shape, it is left to float while cooling after an initial bedding in by hand. The sample shape adjustment is then checked to ensure it is regular (i.e., any fringes visible in the grazing incidence interferometer are approximately circular) and can then be flattened using the shape adjustment.

There are several other desirable items which can help. A grazing incidence interferometer (Section 2.9) also allows the initial sample surface to be accurately adjusted flat and parallel to the jig outer ring so that there is a good chance of avoiding destroying the device on the first run. An optical flat and monochromatic light allow these operations to be carried out to much higher accuracy later in the process. A good quality microscope with sufficient height to allow the polishing jig access under the objective lens; having a camera to record progressive images each time the sample is checked can save hours of work repeating the test if, at the last minute, the desired layer or group is overshot. Lastly, the use of random sweep of the jig at a frequency around the plate rotation speed is used to enable the jig motion to become an invaluable guide to the behavior of the sample as described next.

After aligning the sample with the drive ring and making sure it is flat, controlling the polishing progress is the next challenge. Lowering the sample onto the plate with the micron indexing feature is the first step. Mark the depth pointer position on the jig top ring with a felt-tip pen for reference.

Lift the sample some 20 μm and then run the plate at about 10 to 15 rpm for a few minutes. Presuming that the plate has been properly conditioned flat and free from ripple, then the jig should rotate at a consistently even speed. It may take a minute or two for the abrasive slurry flow to even out and wet the plate. A jig that does not rotate evenly is a sign that the plate surface (or jig outer ring bottom surface) is out of shape; it is essential that this is corrected before proceeding.

Some basic sums will show that a distorted plate or outer ring surface can easily angle the sample out of true to the plate by more than the 1 second of arc mentioned earlier.

Once the jig is rotating evenly, then lower the sample until it is some 2 μm above the zero mark made above. Start the plate as before and observe the jig motion. As it precesses with sweep across the polishing track, you may see some regular point in the rotation where the jig slows as the off center sample contact retards the jig rotation. This is promoted by the design of the jig and alone permits the accurate contact point of the sample with the plate to be found and recorded by marking the jig as above, or by recording the reading with marker pen on the micron indexing feature.

To some extent the tendency for any polishing track to wear slightly convex and toroidal over time will lead to a situation where the sample only contacts the plate as it moves inwards with sweep against the direction of rotation. This variation of plate shape over the width of the polishing track should periodically be checked with the gauge. As the gauge is moved along a plate radius, the reading will vary, showing usually a minimum at the center. If the variation is less than about 3 μm and the readings are the same at several radii around the plate circumference, then this is not a fault but a feature which can be used to achieve a more predictable stock removal and surface accuracy as described in the following paragraphs.

If no sample contact is observed and the jig rotates evenly, lower the sample 1 μm and repeat until the periodic slowing of rotation is just apparent. This is now the true zero and if the sample is checked under the microscope it should be possible to see the point on the sample at which contact with the plate surface has been made. If this is neither in the center of the sample nor has resulted in an even contact pattern, then use either the grazing incidence interferometer or optical flat to correct the sample alignment and flatness (Section 4.4). It will then be necessary to re-establish the zero point as above. It should only have moved a micron or two.

Once even contact has been achieved, it may be necessary to run up to several minutes, checking the surface under the microscope at least each minute, until signs of breakthrough into the next layer are observed.

The top layer of devices is often hard silicon nitride as a protection. In this case, if progress is not being made, then a change to cerium oxide abrasive can be made, as the acid nature of this will help to remove the silicon nitride layer. The alternative is 0.25-μm polycrystalline diamond, but in the absence of a spare plate this is something of a last resort, as it will embed the plate surface with diamond, making it perhaps difficult to control the stock removal rate later during removal of softer layers.

As material is removed, the load on the sample while it is in contact with the plate will reduce as a higher proportion of the load is taken by the jig outer ring, and it will be necessary to lower the sample one-half or 1 μm at a time to reestablish the jig motion pattern, depending on how many layers have to be removed. In Section 2.3, you can see that the sample will naturally sink into the surface of a composite plate some 4 to 6 μm and the sample load will increase progressively over this distance from zero up to the weight of the mobile parts of the jig (approximately 2.5 kg). This can be used to assess and vary the approximate sample load during contact. Less load is applied if the sample height is set 1 μm below the contact point than if it is set 4 μm below because of the compressibility of the plate surface. Similarly the duration of the contact is less at 1 μm than 4.

This pattern of jig rotation and sweep, combined with regular slowing of the jig due to sample contact with the plate at the point where its eccentric progress crosses the centerline of the polishing track, allows the operator to develop a feel for the way in which layers are being removed. With practice, the number of revolutions to remove a specific type and thickness of layer can be predicted and confirmed by regular visual checks of the surface.

In this way the process is not dissimilar to the hand operation, except for the vital factor that the angle of the sample to the plate is very accurately controlled. As at the start, during each observation, if there is sign of a new layer being revealed (e.g., at one corner as on Figure 9.1) then corrective action can be taken on the jig to alter the angle or flatness of the sample.

Semiconductor device delamination is then one of the most challenging of applications for lapping and polishing technology. An attempt to start this without adequate preparation is a sure recipe for disaster and it may pay to review the first half (especially Chapter 2) of this book before risking destroying an extremely important or expensive device which has been submitted for investigation.

References

[1] www.americanepoxy.com/g10fr4sheets.html.
[2] www.tufnol.com/tufnol/default.as.

10

Metallurgical Polishing and Microscopy

10.1 Processing

There are some fundamental differences between polishing for metallurgical examination and for the semiconductor and optical work considered up to this chapter.

- Samples are usually of irregular shape and are often encapsulated, or vacuum impregnated, in resin for preparation.
- Most examination requires a cross-section of material, including any surface coatings, rather than a slice parallel to the surface.
- Emphasis is on fast preparation and scratch-free finish rather than on precision and flatness.
- Most materials, including ceramics, are granular in nature with distinct boundaries, which can cause local lapping

during attempts to polish. For this reason, fixed rather than loose abrasive methods are predominant. For the same reason diamond abrasives are common, as they produce improved planarization of a surface with grains or areas of varying hardness.

- Where precision flatness is required, samples prepared by these methods may be finished on soft metal (tin or lead) plates.
- Automatic preparation stages where blocks of several encapsulated samples are simultaneously processed are available, but many samples are individually prepared by hand on small rotating disk polishers.

10.1.1 Process Stages

Multistage processes are common, for instance, three sequential silicon-carbide-fixed abrasive grinding steps using progressively finer abrasive papers, followed by coarse (often 6 μm) and fine (3- or 1-μm polycrystalline diamond stages). Recently developed disposable rigid composite discs give improved durability and performance over silicon carbide papers. Final stages tend to use silk type polishing pads with loose abrasive [1].

10.2 Examination

The purpose of metallurgical sample preparation (rather than the production of an atomically flat surface) is to produce a surface for microscopic examination which emphasizes (or reveals) minute differences in surface reflectivity and image contrast, which assist visual identification of the full range of features in the material structure. This is aided by the viewer's in-depth knowledge of his or her microscope and the various techniques which are available for enhancing the image contrast, both within the microscope and in preparation of the sample. This is a very specialist field and the reader is referred to the bibliography for a full explanation [2].

Some of these methods, however, can be used to advantage in the examination of planar amorphous or crystalline samples when combined with good practice in the basic microscope setup.

10.3 Microscope Setup

Good sample illumination is the key to microscope images which truly represent what the eye sees directly on a much smaller scale. The first essential is to prevent gross alteration of the colors in the image. If a room is in low color temperature light, such as candlelight at 2,000K, then image colors in a microscope with normal incandescent halogen bulb light (3,000K) will appear very different, and the sample will look different again in a fluorescent light (4,500K) when the sample is removed from the microscope. The simple solution is to use a Xenon bulb (commonly available in 12 volts for car headlights), which has a color temperature of about 5,000K in the microscope's illumination system. As this is very close to natural light, check that the image captured on the microscope's camera appears the same (but much larger than that visible to the naked eye). Most microscope illumination software has the ability to alter the color balance of a digital image.

The illumination system sets out to light the sample evenly, without images of the light source appearing in the microscope field of view. The diaphragm adjustments in the field of view allow adjustment of brightness (resolution) and contrast of the image, which appear directly in the eyepieces.

To obtain the best image in terms of information visible, it is essential to choose an objective lens with the highest *numerical aperture* (NA) available for the magnification required to show the desired field of view. The higher NA allows light from a wider range of angles to enter the objective lens, and this is responsible for improving the information in the image [3].

Working with a low power (10×, for example) objective lens, focus on the sample and close the diaphragm lever or wheel until the diaphragm is closed, showing only a spot of light. If this is not in the center of the view, adjust centering until it is accurate and then open the diaphragm until it is just outside the field of view. This is the optimum initial setting for contrast. Next, use the bulb brightness control to vary the brightness until the optimum detail in the image is visible. Closing the diaphragm slightly may increase the final image contrast a little. These settings should, on a reflected light setup, be adequate for the full range of magnification objectives available [4].

References

[1] www.kemet.co.uk/product.asp?productID=1800&prodCat=Polishing?referrer=google&gclid=CJ31xqO5-agCFQRqfAod4VKQUQ.

[2] Geels, K., "Metallographic and Materialographic Specimen Preparation, Light Microscopy, Image Analysis and Hardness Testing," *ASTM International*, January 2007.

[3] www.microscopyu.com/tutorials/java/objectives/nuaperture/.

[4] Piston, D. W., "Choosing Objective Lenses: The Importance of Numerical Aperture and Magnification in Digital Optical Microscopy," *Biol. Bull.*, Vol. 195, August 1998, p. 14.

11

Laboratory Setup

Previous chapters raise the specter of a long drawn-out, trial and error process, and it is worth thinking out the best lab layout before you start. Clearly, lapping and polishing abrasives do different things and ought to be kept apart: if possible, done in separate rooms—at the very least on separate machines. If you do have to polish on the same machine you used for lapping, then expect long delays cleaning and reworking machines in between operations. Once again, lap to the finest possible finish before the changeover.

11.1 Equipment Locations

Alumina is the main abrasive material for both lapping and polishing (depending on the sample hardness), so it must be kept away from silicon carbide and diamond lapping operations. Very often, sawing operations are housed in the same room as lapping, to maintain the cleanest possible environment in the polishing area.

This is simply a matter of keeping operations that use monocrystalline grit, which resists breakdown into smaller particles, from those involving friable polishing abrasives, in different rooms. The former can be carried to rooms where polishing operations are carried out. A change of clothing between these areas is important; using a simple lab coat only for lapping. A change to more covering clothing can be made for polishing (see Section 7.3). Geological samples, for instance, should be lapped and sawn in one room, but polished in a separate one so that the fine polycrystalline diamond particles can be kept uncontaminated by much larger monocrystalline particles, which may be generated by lapping or sawing. A valuable part-finished section only 15- to 30-μm thick could be destroyed by a single diamond or silicon carbide particle, which could be up to 50 μm in diameter in a laboratory, and up to 10 times this in production environments.

Sinks are essential in all areas, preferably with warm water, and each machine should have at least a 50-liter plastic drain container so that used slurry can be handled and disposed of in such a way as to comply with local regulations.

Figure 11.1 shows some of the equipment used to test and evaluate polished finishes. The cabinet at the rear can house jigs and fixtures. Abrasives should normally be housed in a separate room with its own sink.

11.2 Laboratory Layout and Dimensions

Structures around the machines shown in Figure 11.1 are laminar flow units providing a flow of filtered air for the most exacting work. Although this room has air conditioning to control temperature, it is not to recognized clean room standards and does not have humidity control, which would be essential for processing water-soluble samples.

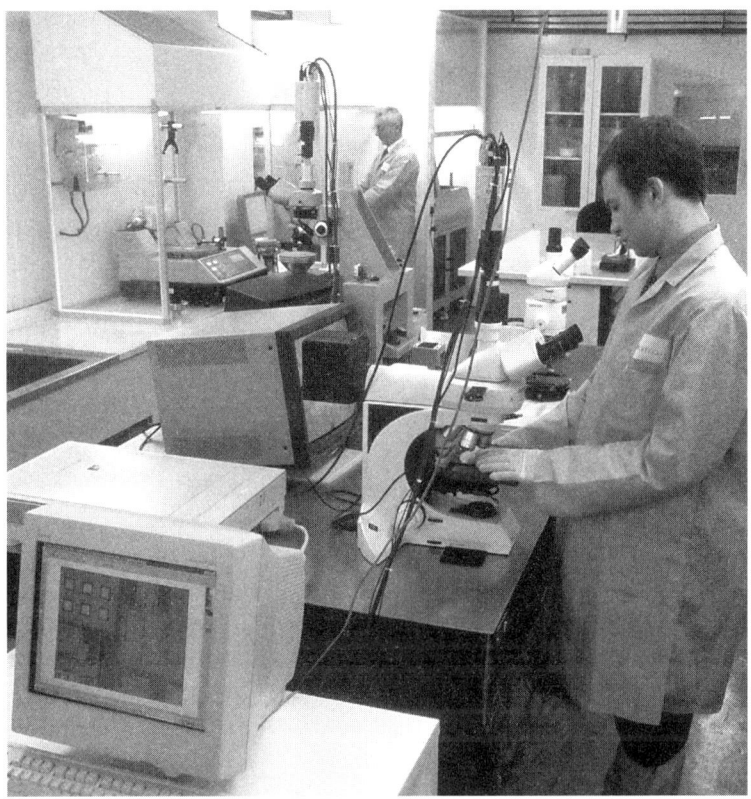

Figure 11.1 General view of a well-planned lab. The machines have 12-inch diameter plates, the size of which determines the dimensions and layout of the lab.

The lapping and polishing machine (Figure 11.2) itself has several basic functions, not simply rotation of the plate: if you cannot find the following features on the machine you have available then think of introducing them at the start.

1. Effective control of plate speed down to 5 rpm or so from 70 rpm.

2. Mounting pillars or jig control arms with at least one which can rotate to provide jig sweep radially across the

working area at a speed independent of the plate rotation speed.

3. Mechanism for mixing slurry to keep abrasive in suspension and deliver a continuously variable flow to the plate (down to as low as one drop per 30 seconds).

4. Easy to clean work area that is waterproof and has a durable finish.

5. Simple mechanism to allow removal and exchange of plates.

6. Most important, a plate diameter which is at least two and one-half times that of the largest sample, block, or fixture you intend to use.

Figure 11.2 A high-quality lapping and polishing machine. The plate diameter is 12 inches.

Each of these functions individually will have an impact on the room layout and forethought is required. Creating a room without this attention to detail will double or triple the amount of work required to perform both process and maintenance.

Such a machine (Figure 11.2) is approximately 2 ft/sq, and so to allow for essential room at the side of each machine for placing jigs, handling plates, and auxiliary equipment like vacuum system components, a length of worktop of about 4 ft long is necessary for each machine. Standard 2 ft wide worktop is not wide enough, and 2-ft 6-in material should be used.

Room layout is usually dictated by entry points (windows, etc.) but one common feature should be considered. Opposite the machines about 2 ft 6 in away to allow machine access (a lifting trolley, often called a scissor lift, is almost an essential), there should also be a strong table capable of supporting instrumentation, surface plates, microscopes, and so forth. Applications, as detailed in Chapters 9 and 10, indicate that constant access to these is essential. Computers, desks, and other equipment can be placed further away, as can the sink. Workstations containing sinks seem ideal, but the sinks can remain unused because they are often too small to handle the plate size and the surface area is more useful.

One of the most successful labs in the writer's memory had no windows. Its success can be put down partially to the lack of distraction this affords and the freedom from varying light and temperature conditions.

11.3 Optimizing the Process Route

One of the most frustrating aspects of working in a sample lab is having to wait for a machine. The number of machines is always going to be budget limited, and increasing the sample throughput is the best way of achieving budgetary targets. At the risk of repetition, the best route to this is:

- Lap to at maximum 300-nm Ra.
- Use at least one short smoothing or fine lapping stage to reduce this below 100 nm Ra, preferably 50 nm.
- If the first polishing run does not reduce the finish near to that required and the prepolished surface is still visible, do not continue polishing, as the surface form will only deteriorate. Consider repeating the smoothing stage and using a more aggressive polish.
- Continuing to polish when the jig gauge and sample surface show no stock removal in half an hour is a waste of time, resources, and causes high blood pressure.
- Exceptions are diamond and materials with grain boundaries, which are tackled in Section 5.2 and Chapter 10.

11.4 Sample Cleaning

Every time a sample is examined under the microscope it will have to be cleaned. Usually if it is still on a jig it may be sufficient to wipe the sample itself with a tissue and *isopropyl alcohol* or *acetone*. If it is on a mounting plate, then it is usually easier to run it under a tap with warm water after applying a small amount of detergent and rinse, which will ensure the mounting plate surface is clean for remounting. Then, if it is still sufficiently dirty to interfere with the microscope image, this can be followed by the solvent wipe.

The theme of two separate operations runs through most laboratory sample cleaning: it is rarely sufficient to use one method only. In semiconductor fabrication a cleaning line is usually employed, with several baths of different agents following each other. However, for less rigorous applications, wetting the surface with detergent liquid, agitating with a soft brush, and then rinsing with warm water is usually sufficient. Refinements can be added. Use of a commercial filtered and de-ionized water supply for final rinse can virtually eliminate smears on the surface, while blowing the surface dry with pure, bottled nitrogen ensures that no contaminants are added from the local compressed air system.

At least this is a good procedure to follow-up with some form of solvent cleaning. Large volumes of work will justify the purchase of commercial cleaning baths but for a few off samples, which need to be clean enough for subsequent processing under vacuum, the simplest method is a vapor still. Here, the samples are suspended above a boiling vapor which condenses on the surface and washes matter and contaminants down into the boiling fluid. This is a very good method for removal of bonding wax residue. However, use of the solvents mentioned above is restricted and so must involve the expense of a proper fume cabinet and vapor control measures meeting local regulations. Fortunately, modern cleaning fluids are now available which boil between 50°C and 80°C [1] and make the task easier.

For use with these, a simple vapor still consists of a 5-liter (1 gallon) beaker on a temperature-controlled hotplate with half a liter of fluid in the bottom. The top of the beaker is covered (not sealed) with a dished glass so that condensed fluid drops from the center into the boiling liquid. To prevent fluid wastage, the top 75 mm (3 inches) of the beaker is surrounded by a copper strip which a copper tubing coil is brazed to, some 8 mm in diameter carrying cool water, to condense vapor as it reaches the top of the beaker. Work for cleaning is suspended from a hook at the side of the beaker, some two in above the boiling fluid. The whole is contained within a fume cabinet, vented to comply with local and national regulations.

11.5 Safety Regulations

In the United Kingdom, and in many other countries, local government regulations for the disposal of waste material and fluids are supplemented by national legislation. A flavor of this can be gained from the introductory paragraph from the U.K. COSHH Web site. Once again, this stresses the forethought and planning which, when carried out before building a lab, can save much time later [2]:

> COSHH is the law that requires employers to control substances that are hazardous to health. You can prevent or reduce workers' exposure to hazardous substances by:
> - Finding out what the health hazards are;
> - Deciding how to prevent harm to health (risk assessment);
> - Providing control measures to reduce harm to health;
> - Making sure they are used;
> - Keeping all control measures in good working order;
> - Providing information, instruction and training for employees and others;
> - Providing monitoring and health surveillance in appropriate cases;
> - Planning for emergencies.
>
> Most businesses use substances, or products that are mixtures of substances. Some processes create substances. These could cause harm to employees, contractors and other people.
>
> Sometimes substances are easily recognised as harmful. Common substances such as paint, bleach or dust from natural materials may also be harmful.

In the writer's experience, the most useful way of maintaining awareness is the provision of simple notice boards with up-to-date posters that cannot be avoided in the way that e-mail communication can.

11.6 Lab Environment

Several factors in the building supporting the lab are important. For the best work, the room should be air conditioned and, for polishing, air particulate controlled to Class 10,000, although this is rare, except in the semiconductor industry.

Just as lapping operations are kept separate from polishing, then chemical polishing should be in a further separate area with fume extract. This needs to be well designed, as it might well conflict with temperature control and air conditioning. Larger chemical machines use high loads. There may be need for a three-phase electrical supply, as well as compressed air, and facilities for disposal of used fluids and slurry to meet local regulations. In particular, the flooring needs attention so that it is easy to keep clean and does not generate particles. Chemical labs also benefit from piped supply of pure nitrogen gas for final dust-off of semiconductor wafers.

All areas need thought as to the operators' clothing. Normally a simple, regularly cleaned lab coat is sufficient, but chemical areas need full-covering suits which can also be disposable. Disposable paper/plastic overshoes are often put on at the entrance to polishing labs and a simple bench across the door aperture serves to remind operators of the need to use them.

Access to a good lathe, capable of swinging the plate diameter in use, and access to a large lapping machine to prepare the plate surfaces, is also essential.

11.7 Consumables

Every time a sample is checked, there is a demand for wipes and solvent, plus occasionally a simple detergent to clean mounting surfaces under the tap and disposable paper towels. The frustration and delay caused by their absence has consequences way beyond their value. Ensuring they are there is the first job before even unwrapping a new work-piece or sample. The same goes for wrapping tissue and boxing for finished work. An absence of pens, felt tips, and stainless steel razor and scalpel blades is equally disrupting. Lab printers usually run out of paper because they have been purloined for notepaper!

If the most useful item had to be selected by vote, then the author would go for a dozen half-liter flexible bottles with directable plastic spouts for handling cleaning fluids and keeping them away from sensitive items, such as computers. Time saved in not having to go to the sink for every cleaning operation can be massive. Using strong, lined disposal bins for trash is essential.

References

[1] http://solutions.3m.com/wps/portal/3M/en_US/electronics/home/productsandservices/products/Novec_Precision_Cleaning_Fluids/Light/.

[2] www.hse.gov.uk/coshh/basics/whatiscoshh.htm.

12

Using Interferometry

12.1 Basic Principles

An image such as Figure 4.4 is invaluable in setting up any sample in the correct orientation for finishing, especially when the instrument used can image in process surfaces. An understanding of optical interferometry can be a real help in imaging, aligning, and assessing the surface you are trying to produce.

It is simplest to imagine the interferometer display as a map. However, instead of the contour lines on the map being 100 ft or 100m apart above the Earth's sea level, they are spaced at half the wavelength of whatever light is passing at *normal incidence* through and above the reference surface. (Example images are in Figures 4.4 and 5.2, and these are discussed later in this chapter.)

Visible light ranges approximately from 0.4 (blue) to 0.8 (red) μm in wavelength (see Table 8.1). Laboratory interferometers usually image at the wavelength of the helium neon laser

(0.632 μm), so the contours are approximately 0.3 μm apart above the reference surface.

However, grazing incidence interferometers operate at a contour spacing between 1 and 5 μm (normally 2 μm) and achieve this by angling the incident light at just below the angle at which the collimated beam would be reflected (the incident angle is measured to the surface normal, not the surface itself).

This is a much more practical range as, because the beam hits the sample at an acute angle, the interferometer will image a lapped surface as well as a polished one. (Try viewing a lapped surface which is angled away from you at about 70°: it appears reflective instead of dull gray.)

In addition, although 0.3 μm vertical spacing between fringes is suitable for final alignment purposes on a nearly polished sample, in the early stages of polishing, images can be difficult to find, and at this stage there are usually too many fringes. Thus, 2-μm fringe spacing is significantly more accurate than at first sight, because in interpreting the meaning of the contour map, it is easy to guess or interpolate the spacing to perhaps one-tenth of the distance between contours; interpreting the vertical distance to 1/20 of the wavelength (in this case) to approximately one-fifth micron.

The fringe map is generated by interference between light reflected from a reference surface (normally flat) and that reflected from the sample under test. The simplest configuration is that shown in Figure 4.7 where ambient light, or that falling from a monochromatic source, illuminates both the reference surface (underside of the flat in Figure 4.7) and the sample mounted on the jig below it.

If the two surfaces are separated at all, then the distance traveled by light rays reflected from the sample will be different from the rays reflected from the reference surface. Each ray behaves like a wave. Its intensity varies in a sine curve, so that when the rays meet after reflection the intensities are different and the light seen by the eye (or a camera) varies in intensity over the area viewed, forming the dark and light bands called fringes (which we have called contours). Each fringe represents a contour line one-half of the wavelength of the collimated light further distant from the reference surface.

Various light sources can be used, the main requirement being that the light does not come directly from a point source and, to some extent, consists of parallel rays. The most common example is a tube light, which both throws soft shadows and does not vary significantly in intensity over the area it covers. In this respect, it can be called an extended source; tube lights have problems for viewing interferograms in that they tend to show fringes only in that part of the image showing the reflection of the tube itself [1].

Ambient sunlight is also an extended source because our distance from the sun is so great that the light rays are effectively parallel. In addition, the light, in passing through the reference component, will be diffracted and split into its color components. The resulting multicolor fringe pattern can best be interpreted by an optical worker with many years experience. Fringes with improved contrast and visibility are obtained from a collimated monochromatic light source, such as that from a laser, the most common being the helium neon laser at 632-nm wavelength.

12.2 Analysis of Fringe Patterns

Faced with a pattern such as the one shown in Figure 4.4, where the fringes are largely straight, analysis is a matter first of counting the fringes over the area to be assessed. The part is parallel to the reference flat by this figure times the fringe spacing in the direction defined by a line joining the centers of the fringes. Second, the direction of this slope is decided by pressing one edge of the part at the fringe center (assuming the part is above the reference surface, if below, press the reference flat). If that point is further away from the flat than the opposite edge, then the fringes will run away from the point and become wider spaced; if nearer, they will run towards the point pressed and become closer together. Section 4.4 shows an illustration of this, involving both sample and mounting plate, which can provide information on both flatness and parallelism of the sample.

Section 2.9 illustrates the method of quantifying the convexity or concavity of a fringe pattern, part of which is spherical but largely regular (which is the case with most flat samples under test). Simply decide on the specific area of the part to be assessed, draw a straight line across this area joining the ends of a continuous fringe, and estimate the number of fringes (if possible, to one-tenth fringe) that lie between this fringe and the straight line. Multiply the fringe spacing by this figure to get the number of microns concave or convex. Deciding between concave and convex is achieved by pressing the part in the center of the assessment area and observing whether the fringes increase or decrease in radius. That is, they move towards the outside of the broadly circular pattern (convex) or towards the center (concave).

In Section 9.1, there is an example of the way in which, especially in flat, parallel samples, two sets of superimposed fringes may be visible. Worse still, the patterns may seem to interfere with each other (*Moire* fringes). Mobile circular fringes (*Hadinger* fringes) may also appear and complicate the picture. These unwanted effects can be avoided by coating the side of the sample which is remote from the flat with a film of grease, or if convenient, this side can be lapped.

12.3 Normal and Grazing Incidence

A laser setup is often unsuitable for testing of transparent optics because the very uniform (or coherent) nature of the laser beam can lead to multiple reflections, and sophisticated design is necessary in purpose built systems for testing spherical optics. For this reason, older, more conventional extended light sources, such as lights at helium, sodium, and mercury wavelengths, are still used.

However, there have been vast improvements in laser interferometers since the first instruments in 1970. Principle in these is the method of sample alignment which allows almost instant positioning of the sample and finding of the fringe image. Secondly, *plug-in* spherical reference elements allow the interferogram to show only errors in the sample surface, distinct from the multiple fringes representing the radius of curvature. Third, automatic computer analysis of the fringe pattern presents verifiable and achievable results, and the creation of three-dimensional images of the surface under test. Fourth, the video system now used permits zoom, focus, and contrast enhancement of images. Available up to 12-inch aperture, such an instrument is the backbone of a professional optical shop.

However, for rapid sample alignment, successful processing of flat samples and assessment of fine machined or ground, in-process surfaces, this level of sophistication and cost is not required.

If, instead of the collimated light passing through the reference surface at right angles, it passes through at as shallow an angle (A) as possible (say, A = 9.5° to the surface itself), then the fringes are further apart, the effective wavelength L being calculated as follows [2]:

$$L = n/\text{Cos}(90-A) \tag{12.1}$$

where n, the refractive index of the reference flat material = 0.67, then

$$L \sim 4\,\mu\text{m} \quad (12.2)$$

Therefore the fringe spacing is

$$L/2 \sim 2\,\mu\text{m} \quad (12.3)$$

This, illustrated below, is the basic principle of the GI interferometer. The geometry naturally lends itself to using a diode laser whose beam is fan shaped, if the problems of collimating such a beam and presenting it to the reference surface at such a shallow angle can be overcome.

If, as illustrated in Figure 12.1, the collimated laser beam enters a parallel reference component from one side, the normal laws of optics dictate this incident beam will simply be reflected if

Figure 12.1 Principle of grazing incidence.

the angle to the surface normal is greater than B°. B depends on the refractive index n of the glass used. This critical angle B, for a beam traveling from air into glass, is about 50°.

The *critical angle* to the surface normal B for light passing from air into a glass or other transparent medium with refractive index n (approximately 60° for air to glass) is [3]:

$$B = \arc\sin(1/n) \qquad (12.4)$$

To make this illustration physically possible the bottom (entry) face of the glass flat would require a sophisticated and expensive multilayer antireflection coating to persuade the incident beam to enter the reference component. Alternatively, the entry face of the reference component can be angled so that the incident beam enters at less than the critical angle. This effectively turns the component into a prism.

12.4 Introducing the Workshop Interferometer

The Appendix describes and gives access to plans for a very simple, cost-effective, and scalable instrument using the alternative solution. It employs a readily available laser diode module and simple reflective optics to generate the collimated laser beam. The accuracy of collimation is therefore a compromise, and satisfactory contrast of the fringes can only be realized with the sample close to (less than 1-mm distant from) the reference flat. It is, however, a small price to pay for the facility and portability the instrument offers.

In addition, the fringe spacing is variable over the approximate range from 2 μm to 4 μm per fringe. Above this range it emits a secondary beam, approximately collimated, at 670-nm wavelength from the third side of the prism. This is intended for applications such as those shown in Section 4.7 (the effectiveness of this secondary beam and of course the whole instrument depends on how well the finished system is aligned). In normal use, with the reference face of the prism in contact with the test face the image is viewed through this third face, as shown in Figure 12.3.

In workshop situations where a fine finish is required on a level or on a recessed face of a part, then the ability to image both the face after fine machining and the distortion on this face as clamping fixtures are released, can mean that significant machine time is saved. Figure 12.2 shows this scenario.

In lab use, sitting workshop interferometer directly on the outer ring of a jig, the rectangular image format allows contact and imaging of a jig outer ring, plus the sample as shown in Figure 12.3. Time is saved on vacuum mounted samples with this arrangement, as the jig can remain on its stand at the side of the

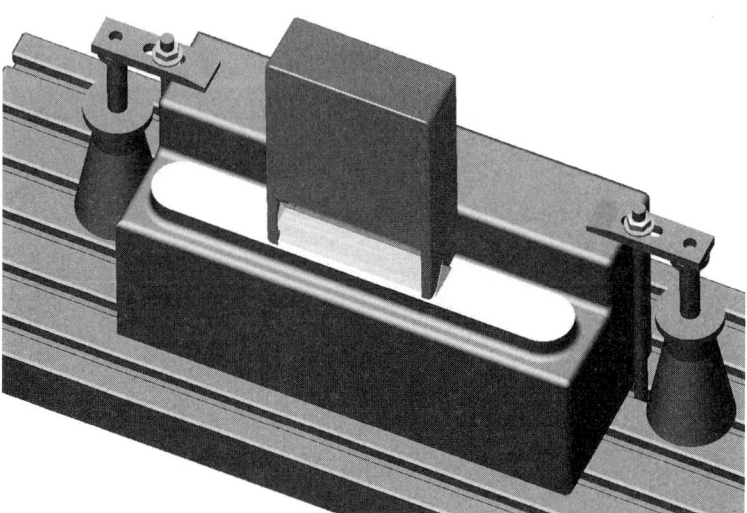

Figure 12.2 Workshop application with machined work-piece clamped to the slide. The interferometer is in place to display the face flatness, and demonstrate the effect of the clamping forces on the fine machined face.

Figure 12.3 Laboratory application showing interferogram of a sample.

polishing or lapping machine. The vacuum connection can still be in place during testing.

It is worth noting here that, using a hard glass for the reference prism, there is very little degrading of the image quality with minor scratching of the glass reference surface and satisfactory use of the instrument can be continued even though the surface has a scratched and damaged finish.

References

[1] Karow, H. H., *Fabrication Methods for Precision Optics*, New York: John Wiley & Sons, 2004.

[2] Alcala Ochoa, N., G. Mendiola, and J. E. A. Landgrave, "Flatness Measurement Using a Grazing Incidence Interferometer," *REVISTA MEXICANA DE FISICA 55 (1) 66-7, Centro de Investigaciones en Optica*, Guanajuato, Mexico, 2005.

[3] Sears F. W., and M. W. Zemansky, *University Physics*, Reading, MA: Addison-Wesley, 1954, p. 734.

Bibliography

Analar Standards for Laboratory Chemicals, 5th ed. revised, Formulated and issued jointly by the British Drug Houses Ltd., and Hopkin and Williams, Ltd., London: Eyre and Spottiswoode, 1957, pp. xvi, 397.

Balmer, R. S., et al., "Chemical Vapor Deposition Synthetic Diamond: Materials, Technology and Applications," in *Journal of Physics: Condensed Matter,* Vol. 2, No. 36, Element Six Ltd., Kings Ride Park, Ascot

Boas, W., *The Mechanism of Electrolytic Polishing*, Oxford, England: Pergamon Press, Ltd., 1959.

Geels, K., *Metallographic and Materialographic Specimen Preparation, Light Microscopy, Image Analysis and Hardness Testing*, ASTM International, January 2007.

Karow, H. H., *Fabrication Methods for Precision Optics*, New York: John Wiley & Sons, Inc., 2004.

Klein, C., "Minerals and Rocks," in *Exercise in Crystallography, Mineralogy, and Hand Specimen Petrology*, New York: John Wiley & Sons, 1998.

Logitech SDG2.

Oberg, E., F. D. Jones, and H. L. Horton, *Machinery's Handbook*, New York: Industrial Press, Inc., 1979.

scholar.lib.vt.edu/theses/available/etd-061899-103951/unrestricted/cain1.pdf.

Sears, F. W., and M. W. Zemansky, *University Physics*, Reading, MA: Addison-Wesley, 1954.

White, C. H., "The Abyssal Versus the Magmatic Theory of Ore Genesis," in *Economic Geology,* Vol. 40, No. 5, August 1945, pp. 353–360.

www.almazoptics.com/LiNbO3.htm.

www.crystran.co.uk/lithium-fluoride-lif.htm.

www.lumonics.com/Materials/H0607_MgF2_Product_Sheet.pdf.

www.taylor-hobson.com/autocollimators.html.

Glossary

Accuracy: The property of a measuring system which relates the closeness of the measured value to the correct value. It should be carefully distinguished from Precision, which is the extent to which a series of consecutive measurements show the same result.

Acetone: One of the most common industrial solvents, acetone is best known as the main constituent in nail varnish removers and paint thinners. In general lab use, it is one of the most effective cleaners, better for instance in cleaning a sample for microscope examination than *isopropyl alcohol.*

Alkaline silica sols: See *Colloidal silica.*

Alumina: Aluminum oxide is the most commonly used abrasive. Not only is it suited to lapping soft rocks and a large number of semiconductor materials, it can be used with either a paper or polyurethane pad for polishing most materials including metals. Available in a large range of sizes from 0.3 μm to 100 μm, there are two forms, calcined and fused. Calcined alumina is a very pure crystalline form and preferred for polishing operations. Fused alumina is less pure and has a greater tendency to clump in lapping operations, causing scratching. The so-called ultrafine alumina can be synthesized rather than milled from larger aggregate, and thus is easier to obtain in uniform particle sizes. Alumina is aluminum oxide, which occurs in many natural forms, the protective coat on aluminum, ruby gem stone, sapphire gem stone, and *corundum* (which does not display the color forming impurities seen in the gem stone forms).

Most alumina forms do not break down very easily into smaller particle sizes and are not quite so popular for fine polishing as polycrystalline diamond, which is highly stressed in manufacture and thus breaks down readily, or cerium oxide. This makes alumina ideal for lapping of softer materials.

Analar: A set of standards for laboratory chemicals first published in 1957 [1].

Autocollimator: An optical instrument used to align two surfaces, or measure the magnitude and direction of the angle between them. As measurement of this angle is often required with extreme accuracy down to 1 sec (1/3600 degree—see Chapter 6) of arc the practical range is usually limited to half or one minute of arc, making "finding" and maintaining the images of the test surfaces very difficult, except for some top-range digital units which have integral sighting eyepieces [2].

Autocollimators operate by projecting a beam of collimated light and detecting the angular deviation between this and a returning beam of light reflected from the surface under test. It thus requires that the surface under test is reflective, as opposed to the grazing incidence interferometer which can align and analyze surfaces which are fine lapped.

Beilby: Sir George Beilby discovered that during polishing the surface of gemstones actually melted and flowed as a "glassy" layer over very fine scratches. He proved it by noting a certain scratch pattern, polishing the surface, and then recovering the scratch pattern by etching away the polished surface with acids.

Birefringence: A very important effect on materials which have different characteristics along different axes is that the axes have different *refractive indices*. This includes crystals, some polymers, and even homogeneous materials which are under stress. Of particular importance are those which develop this effect in a nonlinear way under applied voltages (electro-optic) or acoustic (acousto-optic) signals. These enable designers to control light passing through the materials in a way analogous to the control of electric current by transistors.

Boron carbide: Is another commercially produced abrasive which is significantly harder than silicon carbide. It is often used as a coarse grit to (30–80 μm) to attempt to recover lapping plates which have been contaminated or glazed by diamond.

Borosilicate glass: The main additive to silica in the manufacture of borosilicate, *glass* is boron oxide. This produces a glass with low thermal expansion and high melting point, and thus finds uses in manufacture of laboratory glassware subject to heating and stress. It is widely used for kitchenware in Europe under the trade name *Pyrex*. The glass plate that forms the top of a vacuum impregnation machine is in borosilicate, as can be seen in the reference flat in the grazing incidence interferometer shown in Figure 2.9.

Boule: Semiconductor crystalline materials are generally grown by pulling a seed crystal vertically from a pool of the same molten material, slowly over a period of days. The resulting rod of material is largely polycrystalline and contains impurities. Full monocrystalline material is then zone refined by lowering a heating coil equally slowly from top to bottom of the boule, drawing a molten zone with it. The resulting monocrystalline material is much purer, as the impurities are drawn to the bottom. Silicon boules up to 22 in in diameter have been grown, and 15 in boules are normal.

Bronze: Copper and tin melted together form an alloy which, in historical terms, has been called bronze. The composition is not precise, but is about 88% copper and 12% tin.

Cadmium telluride: A very effective electro-optic material, cadmium telluride's applications are limited by its toxicity. Initially used for infrared optics and windows, it is now more sought as an electro-optic modulator and for use in solar cell manufacture. The global supply of tellurium may turn out to limit its use.

Calcium fluoride: Occurs naturally as fluorite and is the main source of hydrofluoric acid. It is harder than magnesium fluoride and slightly water soluble, but is also used for ultraviolet to infrared lenses.

Canada balsam: Also called Canada turpentine or balsam of fir, it is a turpentine which is made from the resin of the balsam fir tree. The resin, dissolved in essential oils, is a viscous, sticky, colorless (sometimes yellowish) liquid, that turns to a transparent yellowish mass when the essential oils have been allowed to evaporate.
 Canada balsam is amorphous when dried. It does not crystallize with age, so its optical properties do not deteriorate. It is also occasionally used in bonding geological thin sections. However, it has poor thermal and solvent resistance.
 "Canada balsam, the cementing medium used, is not sufficiently tenacious to overcome the stresses imparted to the surface of the

polished grain by the polishing process, which cause the grains to crackle and curl away from the glass slide. This paper describes two mounting media which largely overcome these stresses. These media are the thermoplastics glycol phthalate and Gelva V 7. Glycol phthalate is best suited to the polishing of thin sections on a cloth lap." [3]

Cerium oxide: Used extensively in glass polishing, ceria breaks down very easily into smaller particles, and its slightly acidic nature makes it an extremely useful polishing compound. It is normally available in 2 sizes, 3 μm and 0.5 μm. The very finest grade is known commercially as Opaline. It has almost completely replaced previous optical polishing compounds, such as iron oxide (rouge). Cerium oxide absorbs carbon dioxide from the atmosphere which, in water, dissolves forming carbonic acid, which is weak and unstable. This is responsible for some of the polishing characteristics.

Chemical vapor deposition: This can be carried out at a range of pressures from atmospheric to extreme vacuum. A mix of gasses and vapors react in a chamber to deposit a solid or crystal on a substrate which is normally a wafer. The type of growth can be from a thin film, to polycrystalline to monocrystalline epitaxial.

Computer numerical control (CNC): Automated control of machining and assembly operations have grown at a huge rate since the 1960s. The original punched-tape machines were rapidly replaced with computer-driven interfaces and now allow complex machining operations to performed on multiaxis machining centers (see also Section 5.3).

Color temperature: The scale of color temperature (degrees Kelvin) has its zero at absolute zero (minus 273° Centigrade) and rises in the same units, so boiling water is at 373K. The color temperature of a type of lighting approximates to the light from a source at this temperature. So candlelight is 1,850K, an incandescent bulb at 2,700K, fluorescent lighting at 4,500K, daylight at 5,000K, and a CRT screen at 9,500K. The color temperature has its main impact in microscopy and photography where it is important to match original and viewing temperatures.

CNC Machine: See *Computer numerical control*

Collimation: The property of a light beam which defines the extent to which all rays in the beam are parallel. The formal definition requires that all rays emanate from a point source, even if they are then made parallel by optical lenses or mirrors. In fact because of their finite size, laser diodes create a minimum spot size which can be estimated from the minimum angle of deviation of the beam. This is in the region of 0.1° degree.

Corrosion: Because of the need to protect processes from iron oxide and other corrosion-generated particles, it is sometimes necessary to add an anticorrosion agent to the polishing fluid in which the abrasive is suspended. This is particularly true in the case of sensitive geological samples which are finished by lapping for microscopic examination. In this case, the simplest option is to add a teaspoon of sodium carbonate to the water in the abrasive delivery system. It will reduce corrosion on cast iron lapping plates, high carbon steel polishing baseplates, and on jigs and fixtures. (The cast iron plates will take on a slightly lighter gray color instead of the normal dark gray or rust tone.) Another compound which has been used is Benzotriazole, present is some commercial corrosion inhibitors. Obviously it is necessary to check whether this sort of procedure is acceptable to the final recipient of the sample!

Corundum: This is the natural form of aluminum oxide which, with various colors, is called *sapphire*, or ruby.

Critical angle: See Formula (Section 12.2) [4].

Czochralsky: A process for creating boules of both crystalline semiconductor materials and metals where a small "seed" crystal of the material is very slowly rotated and drawn out of a pool of the molten material.

CVD diamond: Since the early experiments in 1952, when diamond deposits were obtained from a simple oxy-acetylene torch flame, development of methods using carbon-rich gasses to create diamond films have continued. The current art is quoted in the reference document as [5]:

> Freestanding polycrystalline diamond wafers are routinely manufactured in discs exceeding 100 mm (Baik et al., 1999; Heidinger et al., 2002; Parshin et al., 2004). Thin diamond coatings are available with dimensions exceeding 300 mm.

This simple result conceals the vast development work behind this achievement.

In principle, polycrystalline materials are composed of small single crystals (grains) bound tightly together by a thin disordered interface.

Section 9.2.2 explains the difficulties and possibilities of polishing this type of material. Many applications demand that the mountainous form of the completed diamond film is lapped back to a continuous surface. This may be only 5% of the maximum thickness of the completed growth depth once the peaks have been removed. Methods described in Sections 5.2.1 and 5.2.2 have been used for this purpose.

Deionized water: Ions are molecules which carry an electrical charge and are present in tap water. Examples are ions of sodium, calcium, copper, and iron, which remain from mains filtration or from piping systems. All of these tend to make the water conductive to electricity and may interfere with both test results and the structure of samples which are being processed. Removal is normally by filtration using specialized resins, by electro-deionization where the water is passed between positive and negative electrodes which attract ions of the reverse polarity, or by a combination of both. Deionized water results.

Diamond: Diamond has a density of about 3.5 gms/cc. In spite of its comparative lightness (iron has a density of 8), its hardness and thermal conductivity make it ideal as a cutting medium. Natural, unstressed crystals comprise monocrystalline particles of crystalline carbon which do not readily break down into smaller particles and so are ideal for lapping. These are formed under extreme temperatures and pressures some 100 miles below the earth's surface and brought up by volcanic activity. Manufacture of synthetic crystals emulates these conditions. See *CVD diamond*.

Dispersion: This is the quantity defining the spread of values of refractive index of a transparent material with wavelengths of light incident on the surface. For convenience, it is measured at a specific wavelength: that of helium at 587.56 nm. It is this property which causes natural light to be split into a rainbow of colors when passing through a prism (or raindrop). It can be varied by specific additions to the melted glass before the mix is cooled.

Electroless nickel: Electroless coatings are deposited in a highly controlled chemical environment without the application of an electric current to promote deposition. Normally a nickel phosphorus alloy is deposited in an even film, which may be up to 100-μm (0.004-in) thick. This makes the coating very suitable for subsequent diamond turning to a final shape.

Embedded device: An embedded device is a subsystem within a device, often a discrete element, which carries either programmable or updatable software in order to perform a specific function. It follows a trend over the last 20 years towards both electronic and electrical devices which, instead of resembling mechanical devices which perform one or a defined series of functions, are addressable by other devices to alter their responses or provide specific information using a common language system.

Ethylene glycol: Also called ethane diol, this is a very useful fluid in the polishing lab. It has a soapy feel and is more viscous than water. In geology, it is used as a carrier fluid for alumina abrasive to moderate or soften the lapping action, when lapping soft rock samples. This effect can be used on a range of fragile materials. Commercially, it is widely used as vehicle antifreeze and also in the manufacture of dynamite. It is a base from which many plastics are synthesized.

Fizeau: A Fizeau interferometer is one of the simplest forms. In principle, a light beam emanating from a point source expands to a collimating lens or mirror and then passes through a reference flat normal to the reference surface (this surface may also be spherical allowing assessment of matching spherical components). A test component is suspended just above the reference flat and interference fringes are developed which can be viewed either by a beam-splitter arrangement in the expanding section of the light beam or directly if the sample is transparent.

ELID: When grinding silicon wafers, it is normal to use progressively smaller diamond grit sizes to minimize subsurface damage and the material to be removed in subsequent polishing.

As grit sizes reduce, it becomes more difficult to dress the wheel. Conventional bronze-bound wheels are conductive and an electrolytic method called ELID can be used to progressively remove the matrix allowing the worn abrasive particles to drop out.

Epitaxial: Growth of a single crystal material on a single crystal (monocrystalline) substrate. The *substrate* acts as the seed crystal for the grown material, so this takes on the crystal structure of the substrate.

Ferrite: The purest form of iron, responsible for the magnetic properties of iron and steel.

Fused silica: Fused silica is a manmade, very hard, and transparent glassy, amorphous solid. It has become a favorite for making reference flats and test plates of high quality. Chemically, it is pure silicon dioxide which has been heated through the melting point and then made to cool at a rate sufficiently fast to prevent crystallization.

Fused quartz: It is also silicon dioxide, but with a low level of alumina impurity from its natural source, which is mostly Brazilian quartz sand.

Garnet: Garnets are naturally occurring crystalline silicates frequently appreciated as gemstones because of colors which originate in metal oxides in the crystal structure. They are still used as abrasives for functions, such as water jet cutting, and skilled operatives can prefer them for final finish polishing because of their ability to break down to very small particle sizes. The *Mohs* hardness is approximately 7.

Gallium arsenide: As an IR transmission and semiconductor material, gallium arsenide is now second only to silicon. Its raw materials are readily available as biproducts from aluminum and zinc production. Its use has grown with the demand for diode lasers, and IR optical uses are less common, except for high-power applications where it out-performs zinc selenide. Originally, this was polished using bromine in methanol alone, or with alkaline silica sol additions. Peroxide alkaline etch solutions have since mostly replaced bromine in methanol for this purpose.

Germanium: This is a metal found only as a trace in deposits of other metals. It thus becomes available in small quantities only largely as a biproduct of their extraction. Its relative hardness allows it to be easily polished, and it has been diamond machined. Its brittle nature makes it somewhat liable to impact damage, but its high strength allows use as windows in challenging environments. Its performance at high temperatures is impaired because the refractive index increases rapidly with increasing temperatures above 50°C, making the optical system in which it is mounted less efficient with subsequent generation of local heat and possible self-destruction. It is chemically very stable and thus normally polished by mechanical methods.

Glass: See Section 8.4, *Silica, Pyrex*. Glass is formed when the cooling rate of molten silica is controlled at such a rate as to inhibit crystallization. The properties of the glass depend on this rate and, more markedly, on additives to the silica which significantly change the glassification temperature. Glass is an amorphous solid, that is one where the molecules have no arranged order or alignment. Many different materials exhibit partial glassification, particularly *polymers*.

Glycol phthalate: Prepared from ethylene glycol and phthalic anhydride, glycol phthalate is a clear solid, which starts to soften at

70°C and is fully molten at 120°C. Its refractive index is similar to glass, and thus it can be used as a bonding agent for preparation of very thin sections of materials which will be subject to microscopic analysis. In this, it has largely been superseded by epoxy resins, in cases where it is not necessary to release the section from its bond after testing. The wide melting range makes it ideal for destressing bonded samples as the bond can be heated to a point where the sample relaxes without releasing from the mounting plate.

Hadinger fringes: These appear in optical test systems and represent lines of equal inclination between plane surfaces which are under monochromatic light. They can be identified for elimination purposes by the fact that they move when the observer's head moves.

Incident light: In optical systems, light falling on a surface is called incident light. The angle of incidence is not the angle at which the light strikes the surface but the angle to the surface normal.

Infrared: The infrared wavelength region extends from red at 0.7 μm to 300 μm and covers those wavelengths commonly known as heat, or more accurately, thermal radiation. ISO 20473 divides this into:

Near Infrared from 0.78 to 3 μm, the lower values being used in fiber optic communications (silica glass fibers transmit particularly well in this region).

Mid Infrared from 3 to 50 μm. Used in heat-seeking missile applications.

Far Infrared from 50 to 1,000 μm. Mainly used in thermal imaging.

Isopropyl alcohol: Isopropyl alcohol is best known as the active component in aerosol car windscreen de-icers. As such, it is one of the least toxic of laboratory solvents, although many labs have replaced it with biodegradable citrus-based fluids. The vapor is highly flammable.

KDP: Highly water soluble electro-optic crystal. It is very sensitive to temperature fluctuations and mechanical shock. However, crystals have been grown to over 15-in across.

Kurtosis: Kurtosis is a measure of the randomness of heights and of the sharpness of a surface. A perfectly random surface has a value of 3; the farther the result is from 3, the less random and more repetitive the surface is. Surfaces with spikes are higher values; bumpy surfaces are lower.

Laser diode module: Specifically, this refers to a metal capsule containing both a laser chip and the relevant power supply stabilizer, ensuring that a range of voltages can be applied without causing damage to the chip. Often, there is a collimating lens with focus adjustment allowing the laser beam to collimate to a minimum diameter beam. A parallel, or screwed, outside diameter allows for simple fixing and rotation of the module.

Lithium fluoride: Lithium fluoride is a very stable salt with unsurpassed transmission at the ultaviolet end of the visible spectrum. It is thus used for optics in high-resolution semiconductor lithography and x-ray applications. Its water solubility is 0.27% [6].

Lithium niobate: A 3-in diameter boule of lithium niobate is shown being sawn into wafers in Figure 6.1, and a stack of prepared slices ready for edge polishing in Figure 5.5. It is a nonwater-soluble [7], transparent crystal with excellent polishing characteristics using alkaline silica sols. Its electro-optic characteristics make it the ideal substrate for device construction and research.

Live weight: When a jig is used, the weight of all moving components must be subtracted from weights added to provide the calculated load on the sample. The easiest way to do this is use a fixture which supports the outer ring and then measures the weight of the sample plus all components attached to it. If there is a load control mechanism on the jig, then this can be adjusted and the resulting effect shown on the fixture.

Magnesium fluoride: This is commonly used for antireflection coatings on optics over a wide range of wavelengths. It occurs naturally as crystalline sellaite. It is relatively soft with a *Mohs* hardness of 5 to 6. This makes it suitable for machining with diamond tooling, and is manufactured in diameters up to 150 mm for lenses, windows, and domes in the infrared region [8].

Modulation: This is the process of altering properties of a carrier signal (one which can be transmitted) so that digital or analog information can be carried by the signal. For instance, an amplitude modulated radio signal varies in strength at the frequency of the modulating signal whereas a frequency modulated signal varies in frequency at the frequency of the information. Then the radio receiver either extracts the amplitude information (AM) or the frequency information (FM) to present the transmitted program [9]. This process can only occur if the modulating device has a *nonlinear* characteristic (otherwise, any signal superimposed on it will simply add or subtract amplitude or frequency, not multiply, which is the function required to carry the modulating signal). Modulation can be made to occur at any of the frequencies shown in Table 8.1. (The radio analogy is simply an example.) It is very important in the context of this book, as it provides the reason for creation of the electro-optic crystals introduced in Section 5.4 and the incentive to find ways of processing them.

Mohs hardness scale:

Talc	1	Very easily scratched by the fingernail
Gypsum	2	Can be scratched by the fingernail
Calcite	3	Scratched by a copper penny
Fluorite	4	Very easily scratched with a knife
Apatite	5	Scratched with a knife with difficulty
Glass	5.5	
Orthoclase	6	(Hard, crystalline natural mineral)
Steel needle	6.5	Scratches glass with difficulty
Quartz	7	Scratches glass easily
Topaz	8	Scratches glass very easily
Corundum	9	Cuts glass
Diamond	10	

[10]

Moire fringes: Where there are two superimposed sets of lies in a view (for example, interference fringes—equal distance contours and *Hadinger* fringes—contours of equal inclination), then the patterns themselves in the image can interfere with each other, forming a third set of light and dark bands; often this leads to confusion and the inability to interpret a simple interference pattern. The first recourse is to eliminate the Hadinger fringes by coating the top face (i.e., the face remote from the reference side of a transparent sample or reference flat with grease). In the specific case where the fringes occur by interference on the display itself, then a software filter (called descreen) can be used to remove the effect.

Nonlinear applications: See *Modulation*

Outgassing: Porous or friable materials, like earth or coal, need to be impregnated with epoxy resin before a thin section can be prepared for microscope analysis. Damp materials must first be dried, often for days, on a hotplate. Then they must be outgassed so the resin can be drawn into the pores and cracks. This involves holding the material in a close-fitting dish, under vacuum for sometimes a long period until the level of vacuum reaches a preset value. The material is then covered in premixed epoxy and the vacuum released so the resin is forced into the pores of the sample (see Section 5.1).

pH: A scale ranging from 1 to 14, from strongly acid to neutral (7) to strongly alkaline. Water is generally pH 7, although local variations occur, *deionized* water being the most reliably close to 7. Tests are easily conducted using commercial test papers, which change color at a specific pH range; however, more accurate measurements, which are required for some types of chemical polishing, are conducted using pH meters which generally require frequent calibration with standard solutions.

Plug-in optical elements: In high-quality optical interferometers, reference elements which may be flat or spherical are arranged on sophisticated bayonet-type fittings, so that their location is fixed and repeatable relative to the main optical path.

Polymer: The word polymer refers originally to a long chain molecule rather than a class of materials. Thus in nature, proteins, rubbers, amber, shellac, and silk are examples of materials having polymeric molecules. Synthetic polymers are perhaps better known because of their very wide range of properties, both *thermoplastic* and *thermosetting*.

Poromeric: A specific type of multilayer and nap synthetic polishing surface extensively used in semiconductor wafer polishing. Poromeric was coined by DuPont from the terms microporous and polymeric, and initially described a synthetic leather for shoes (see Section 8.3).

Pyrex: Pyrex is often quoted as a glass type, however it is in fact a trade name. Up until the 1980s, the material used for the popular ovenware was *borosilicate glass*. Since then, soda lime glass has been used, tempered to improve thermal resistance. Some web sources say that the later glass is less durable under thermal stress.

Refractive index: Light passing from a vacuum into a transparent solid is slowed down by an amount defined by this quantity N. It is a constant for a specific material and is defined by Snell's law relating the angle of the light to the surface normal in the vacuum Av to the angle to the surface normal in the material An that is [11],

$$N = \sin Av / \sin An$$

Resolution: A property of a measuring device representing the smallest change a sensor can detect in the quantity that it is measuring. It should be carefully distinguished from *Accuracy* and *Precision*.

Rust: See *Corrosion*.

Sapphire: *Corundum* with trace metal oxides, which can impart colors other than pink or red (which forms are called ruby). Both polycrystalline and monocrystalline forms can be artificially produced. The polycrystalline form is hot pressed (sintered), whereas single crystals can be produced up to 15 in in diameter by the *Czochralski* process. Monocrystalline wafers are much in demand for semiconductor process substrates.

Silicon carbide: Silicon carbide is the polycrystalline compound of silicon and carbon. In contrast to polycrystalline diamond, its grains are very tough and less dense. It is therefore rarely used for polishing, but widely used for lapping rock samples for thin sections in a range of sizes from 9 to 20 µm. A common trade name is Carborundum. It can be sintered into various shapes and is an ideal substrate material for advanced semiconductor work because of its stability. As a compound, it is very rare in nature. There is room for confusion with common or trade names here (for instance, there is also carbonado, a natural polycrystalline diamond that is black in color) so it is wiser to stick to the full name for abrasives! Hardness on the Mohs scale is about 9.3.

Silica: See *Silicon dioxide*. Silicon compounds form a large proportion of the Earth's covering.

Silicon dioxide: A hard crystalline material most commonly known as sand or silica, the crystalline form of glass. In combination with metal oxides it forms many of the less valuable gem stones such as *garnet*. In semiconductor layer construction, it is the hard oxide layer used to provide stability and insulation between active layers or layer groups, and can be very difficult to remove when encountered during delayering of a device. It is the abrasive component in *colloidal silica*, found there in particle sizes of about 1/20 micron, or two millionths of an inch.

Silicon nitride: Also used as an insulating layer multilayer devices and less hard than the carbide, silicon nitride is a manmade material, which is particularly tough and wear resistant. It is commonly used as a barrier to oxidation. In device delayering, it can be removed either by very fine polycrystalline diamond abrasive or by extended polishing using cerium oxide.

Smoothing: See Section 3.4.2. Conditioning of smoothing pads: The author has successfully used smoothing pads up to 1m in diameter in resin material, when it was found to be relatively easy to run a separate conditioning ring on the pad and temporarily apply loose alumina abrasive about one-third the size of the diamond particles. This effectively conditioned the surface and gave the ability to control the pad surface shape. It was then possible to flood the pad surface with water and use a brush to ensure that none of the loose residue remained. Because of the diamond impregnation, this method will effectively remove hard (e.g., silicon nitride) coatings on device wafers.

Substrate: In the context of this book, substrate always refers to a wafer or block on which the sample for processing is mounted, not the sample itself.

Thermoplastic: Thermo-softening, as opposed to thermo-setting *polymeric* material. Thermosetting plastics can be heated to above their melting point and reshaped on cooling. One of the earliest was celluloid. Important current examples are acrylic, nylon, polyethylene, polypropylene, polycarbonate (often used for plastic lenses), and polyvinyl chloride (PVC).

Thermoset: On heating, thermosetting materials develop cross-links between the long chain molecules which define them as *polymers*, which unlike *thermoplastics*, solidify instead of melting. The first commercially successful thermosetting polymer was Bakelite. Important modern thermosets include both melamine and epoxy resins.

Tracking: Tracking of a lapping or polishing plate occurs when a hard sample wears a toroidal area of the lapping or polishing track more than the rest. It can be detected by careful use of a gauge (gauges are available with an extra dial gauge offset towards the outside diameter to facilitate this) [12]. Avoiding tracking can be either by offsetting the sample towards the outside diameter of the jig, using appropriately large amounts of sweep (so the sample covers the whole track), or by running a heavily loaded test block on the same machine. Often a combination of all three of these must be used, for example when processing diamond. A tracked plate which has been used on diamond can only be recovered by machining off the tracked area in a lathe.

Undercut: A method of lapping in which the sample is held rigidly, initially protruding slightly below the surface of the outer ring of a jig. At first it is subject to the full weight of the jig, but the surface laps to above the outer ring, which by then is in full contact with the plate. The load reduces to zero and the abrasive then continues to generate a clearance between the plate and sample. This is the order of 3 times the abrasive diameter, depending on the specific grit used. The advantages of this method are that the parallelism of the sample face and jig mounting face is rigidly maintained, and the surface finish produced is finer than if the sample was in direct contact with the grit on the plate.

Vacuum impregnation: See *Outgassing*.

References

[1] *Analar Standards for Laboratory Chemicals*, 5th ed. Revised. Formulated and issued jointly by the British Drug Houses Ltd., and Hopkin and Williams, Ltd., London: Eyre and Spottiswoode, 1957, pp. xvi, 397.

[2] www.taylor-hobson.com/autocollimators.html.

[3] *Economic Geology;* Vol. 40, No. 5, August 1945, pp. 353-360.

[4] Sears F. W., and M. W. Zemansky, University Physics, Reading, MA: Addison-Wesley, 1954, p. 734.

[5] Balmer, R. S., et al., "Chemical vapor deposition synthetic diamond: materials, technology and applications," in *Journal of Physics: Condensed Matter,* Vol. 21, N. 36, Element Six Ltd., Kings Ride Park, Ascot.

[6] www.crystran.co.uk/lithium-fluoride-lif.htm.

[7] www.almazoptics.com/LiNbO3.htm.

[8] www.lumonics.com/Materials/H0607_MgF2_Product_Sheet.pdf.

[9] scholar.lib.vt.edu/theses/available/etd-061899-103951/unrestricted/cain1.pdf.

[10] Klein, C, Minerals and Rocks: "Exercise in Crystallography, Mineralogy, and Hand Specimen Petrology," John Wiley & Sons, Inc., 1998.

[11] Sears F. W., and M. W. Zemansky, *University Physics*, Reading, MA: Addison-Wesley, 1954, p. 732.

[12] Logitech SDG2.

Appendix: The Workshop Grazing Incidence Interferometer

This appendix contains build and setup instructions for the simple workshop interferometer described in Chapter 12. Now that higher power diode laser modules are available, this design can be scaled, and units can be built with a reference face several times the base 30 × 140-mm size, limited mainly by the cost and availability of the reference prism.

A.1 Optical Path

Two distinct views of the design are shown. Figure A.1 shows the side-view of the interferometer with hardware removed so the beam path is clear. Figure A.2 shows an isometric view designed to illustrate the beam expansion in elevation at right angles to the side view.

The key to both views is as follows:

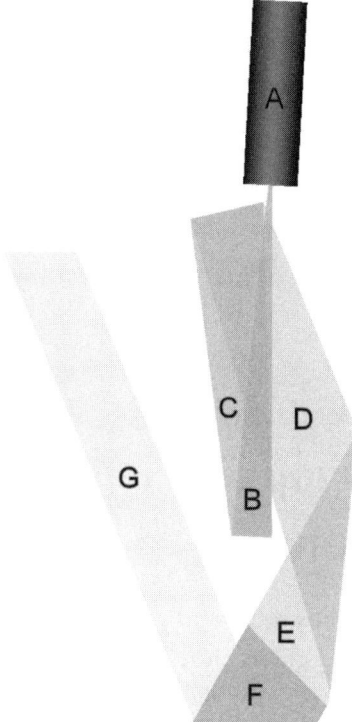

Figure A.1 Side-view of optical path.

- A is the laser module. In this case, a 670-nm diode unit with in-built voltage stabilizer is used. After removal of the in-built collimation lens, the output beam expands at an included angle of 4° in the side-view plane shown, and at 40° in the elevation view.
- B is the first expansion stage from the laser aperture to the first collimating mirror, which reduces the elevation view expansion angle from 40° to 20° without affecting the side view.
- C is the second expansion stage from the first collimating mirror (B) to the second (C). This reduces the elevation

Appendix: The Workshop Grazing Incidence Interferometer   183

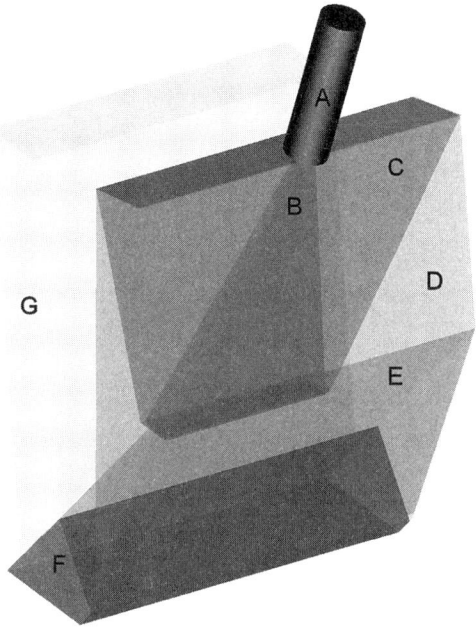

Figure A.2 Isometric view showing beam expansion in elevation in two stages, B and C. Following stage C, the beam is parallel in elevation and continues to expand in end view through stage D.

view expansion angle to zero. Note that the side-view expansion continues at 4°.

- D is the final expansion stage up to the end view collimation mirror (D), which is pivoted at the bottom to provide resolution adjustment.
- E is the final collimated beam, incident on the first face of the reference prism F. The bottom face of F is the reference flat surface.
- G is the collimated beam which emerges from the third face of the prism due to total internal reflection at the bottom reference face, when the resolution control is adjusted to its minimum setting.

In regards to eye safety, as the beam is expanding freely the laser intensity is very low, but operators should avoid looking directly into the beam at all times. In particular, the beam should never be viewed using any kind of optical instrument. Adjustment should be performed in a slightly darkened room, with laser intensity at the lowest possible setting to give acceptable visibility. The beam of the module should never be viewed directly prior to removing the collimating lens with which it is delivered.

A.2 Stage Mirrors and Adjustment

This section explains the function of the various beam manipulation components.

- *Expansion stage B:* This is the natural expansion of the diode laser beam. The module's integral collimating lens is removed *before first attempting to switch on the laser.* This allows the beam to expand fully, limited only by the aperture of the module's inner diameter. It also avoids multiple interferometer images which would be generated by reflections from the beam's transit through the various surfaces of this lens. At the transition from stage B to C, the beam is reflected from the first collimating mirror (B). This is the stage when the rotation of the laser module is adjusted so that the beam accurately fills the profile of mirror B, and the axial position of the module is set to the nominal dimension.
- *First collimating mirror:* This is a strip of high-quality prefinished, stainless steel sheet which is bent to a curvature tighter than the nominal radius necessary to 50% reducing the elevation expansion of the laser beam from 40° to 20°. The strip is then constrained at both ends against a flat aluminum support member so that both the orientation and

expansion angle can be controlled by screws which are loctited [2] in place after adjustment. This adjustment then produces a mirror profile which is approximately parabolic [3]. Setup is comparatively simple, as the beam is simply constrained by the screw adjustment to fill the second expansion-stage mirror (C).

- *Second collimating mirror:* Adjustment of this stage is similar to that of stage B, the beam being focused by the adjustment screws to fill the aperture of the end view collimation mirror (D) at the end of stage D. This reduces the elevation expansion to zero. A simple check is to remove mirror D and allow the beam to impinge on a room wall a few feet from the instrument, when its width in elevation should be the same as that of mirror C. The end view expansion continues at 4°.

- *Expansion stage D mirror:* This mirror is pivoted at the lower end and is adjustable to control the beam incident angle on the reference prism. Its shape will remove the end view expansion, reducing it from 4° to 0 and produces a rectangular parallel beam which is approximately collimated. It may be checked in the same way as mirror C by removing the reference prism and checking that the projected beam maintains the same dimensions up to several feet from the instrument. If the end view expansion is not accurately removed, the parallelism in this plane may be adjusted by moving the laser module along its axis after loosening the clamp. The adjustment, particularly the rotation of the beam at mirror B and C, will then have to be rechecked and then the laser clamp retightened.

Mirror D is in a simple off-axis parabolic mirror to perform the end view collimation, made from a strip of the same stainless steel as mirrors B and C. It is formed in the same way and attached with epoxy resin to a rotating former which facilitates adjustment of the incident angle of the collimated beam on the reference prism. Parabolic mirrors may seem complex, but in use are very simple [3].

- *Reference prism:* Two alternative designs for the prism are provided. The first is for a relatively soft and inexpensive borosilicate glass, the second is harder and more expensive fused silica. The prism angles vary slightly to accommodate the different refractive indexes of these materials.

The designs ensure that the incoming collimated laser beam from mirror D enters the prism as near as possible normal to the first surface to limit reflection of the beam. It then strikes the reference face at such an angle that it passes through it at the correct range of angles (adjusted by mirror D). This varies the fringe spacing of the image of an object placed close to the reference face from approximately 2 to 4 microns. The image is then viewed through the third side of the prism (see Figure 11.3) and may be photographed directly by a digital camera or mobile phone. (The best compromise setting of the laser intensity for this must be established by experiment.)

- *Auxiliary beam:* The laser beam issuing from the reference surface is at approximately 10° to the surface and is thus narrow compared with the beam entering the first face. The mirror D adjustment is arranged so that, when adjusted to above the 4-micron resolution indicator point, the beam striking the inside of the reference face is totally reflected and passes normally out of the third side of the prism. This beam may then be used to illuminate an external reference flat as illustrated in Figure 4.8. The beam intensity is adjusted to minimum in this configuration, unless subsequently overridden by the user during a test.

- *Calibration:* The reference flat configuration lends itself to a simple and visible method of calibration. The only component required is a standard optical flat and a simple rig which, for advantage, can be on a surface table.

The interferometer is clamped upside down to a square block on the table in such a way that the optical flat can be rested on the reference surface protruding to the side of the flat. The test gauge is secured by a clamp in the position shown under the free end of the flat so that this can be raised slowly by turning an M3 grub screw. A 3-mm hex key (visible in Figure A.1) is used so as to force the gauge plunger up against the underside of the flat and raise one end up to create an air wedge between the reference flat and the optical flat. When the interferometer is switched on, a series of parallel fringes will appear between the two reference faces as shown in Figure A.3, as the optical flat is raised at one end. Simultaneously, the gauge can record the amount by which it is raised.

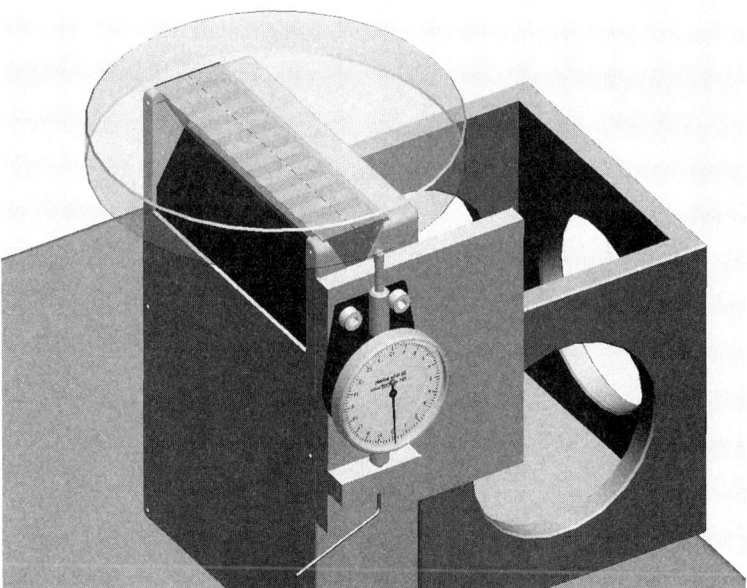

Figure A.3 Test and calibration setup.

If the length L is as shown in Figure A.4 (from the edge of the reference prism to the centerline of the gauge) and the reference prism length is 140 mm, then:

The calibration in *microns per fringe* is:

Change in gauge reading/No. of complete fringes $\times$ 140/L   (A.1)

The test is repeated for progressive positions of the dial control and the results engraved on the casing. Also noted at this stage is the dial position at the point where total internal reflection of the beam occurs and the auxiliary beam G appears.

A.3 Hardware and Detailed Drawings

The list of hardware (bill of material) for building the unit and detailed drawings for manufacture are available at www.artechhouse.com/static/reslib/robertson/roberston.html.

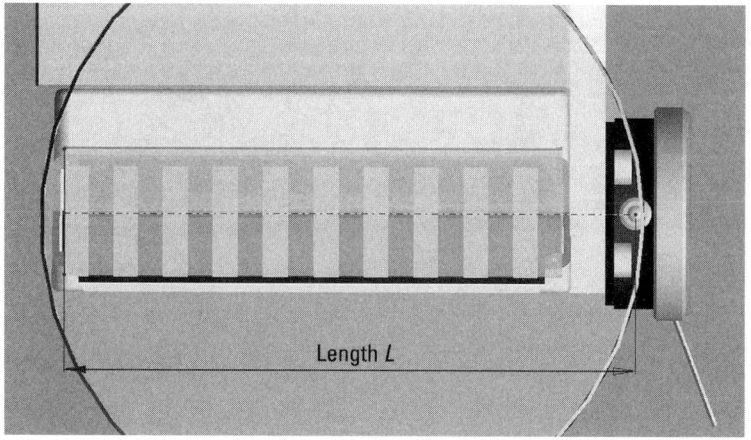

Figure A.4   Definition of length L.

About the Author

Born in Bristol in 1942, Max Robertson was evacuated to a farm in Cornwall until the end of World War II, and then started school in Scotland, where his father went to work for the Naval Construction and Research Establishment. Secondary education was at Daniel Stewart's College in Edinburgh, followed by 4 years in St. Andrews University, where he earned a B.Sc. Mech. Eng.

This was followed by a series of roughly 5-year placements in the motor industry, Ford UK, Ford Werke Ag, Aston Martin, and Rolls Royce, in design and development posts, until the exhaust emission scene emerged to repress his chosen field, which was engine development. A return to Scotland followed, with time in the defense industry focused on diamond machining, and when this came under national economic pressure, Max moved his family to Glasgow to complete his children's education.

Max then acquired 12 years of experience in lapping and polishing, where he started as a development manager for Logitech Ltd. followed by 8 years in machine design for this company. Max is now retired with his long-suffering wife of 40 years, Winifred, and considers himself very fortunate to live within easy traveling distance of his two daughters.

Index

A
Accuracy, 64, 68, 75, 94, 107, 130, 159
abrasive, 19
acetone, 142, 159
AFM, 106,107
alignment, 16, 62,
alkaline silica sols*, 73, 159
alumina, 19, 20, 31, 103, 159
aluminium, 29, 95
angular adjustment, 63, 69, 128
analar, 159, 160
annular, 24, 100, 160
autocollimator, 21,160

B
Beilby, 59, 161

balancing, 34, 46, 94
base plate, 28, 29, 30
beeswax, 38
birefringence, 78, 88, 161
bond, 38, 40, 43, 65
boron carbide, 19, 63, 81, 161
borosilicate, 142, 161, 186
boule, 97, 101, 161
brass, 24, 82, 103
bronze, 30, 31, 98, 101, 162

C
cadmium telluride, 75, 89, 162
calcium fluoride, 89, 162
camera, 129, 135, 149,160
canada balsam, 162
carrier, 41, 71, 167, 171

---
*Terms in italics are defined in the Glossary.

cast iron, 23, 25, 30, 117
cerium oxide, 20, 63, 69, 131, 163, 176
chemical vapor deposition (*See* CVD diamond)
clothing, 145
clumping, 20, 164
CNC, 163
collimated, 69,149,156,164
colloidal silica, 20, 73, 159
color temperature, 163
compressed air, 94, 143
compressibility, 132
conditioning block, 23, 31, 55, 114, 118,121
contour, 43, 103, 147, 149, 173
co-planar, 37, 62, 73
corrosion, 164,
corundum, 165
critical angle, 165
cutoff, 106, 108
CVD diamond, 165
czochralsky, 165

D
de-ionized, 74, 166, 170
de-layering, 28, 67, 128
dial gauge, 22, 26, 34, 176
diamond (machining), 93, 166
diamond (substrate), 82, 83, 166

diamond (tooling), 121, 166
diaphragm, 39, 100
dicing, 100
dispersion, 112, 166
distortion, 17, 29, 36, 39, 43, 86, 122, 154

E
electrolytic, 20, 168
electroless nickel, 94, 166
epitaxial, 15, 82, 163, 168
epoxy, 30, 36, 78, 79, 85, 97,
epoxy bonding, 36, 42
ethylene glycol, 86, 100, 167

F
ferrite, 29, 52, 168
fizeau, 167
fringe, 43, 63

G
gallium arsenide, 70, 89, 169
gauge, 22
garnet, 167
geological, 35, 79, 99, 188
germanium, 89, 169
glass, 23, 36, 78, 126, 169
glycol phthalate, 163, 169
granite, 29, 79, 95, 114
grazing incidence, 16, 43, 62, 64, 78, 103, 129, 131, 181
grinding, 19,25,96,104,118, 168

grooves, 24, 32, 36, 37, 53, 74, 103, 127

H
hotplate, 38, 143, 174
Hadinger fringes, 69, 150, 170, 173

I
infrared, 77, 88, 162, 170
interferometry, 16, 43, 147
isolation, 94
isopropyl alchohol, 159, 170

J
jig, 33, 172, 178

K
KDP, 87, 171

L
laser diodes, 16, 90, 153, 164, 171
lathe, 26, 28, 32, 51, 74, 95, 114, 145
lithium fluoride, 89, 171
lithium niobate, 45, 74, 90, 171
live weight, 50, 172

M
machining, 4, 17, 93
magnesium fluoride, 89, 162, 172

melting point, 39
modulation, 90, 172
Mohs, 173, 176
Moire, 150, 173
monocrystalline, 19, 30, 121

N
nap, 54, 71, 86, 118, 174

O
optical flat, 65, 67, 126, 129, 131
optics, 39
outer ring, 33, 49, 51, 62, 129, 130, 131
outgassing, 174

P
paper, 26, 31, 78, 83, 117, 134, 159
pH, 174
polycrystalline, 20, 83, 131, 88, 131, 138, 162, 165, 176
polymer, 161, 169, 174, 177
polyurethane, 25, 36, 74, 84, 117, 119, 159
poromeric, 174
porous, 36, 79, 117, 174
profiling, 16
Pyrex, 175

Q
quality control, 111, 125

R

radial, 24, 31, 53, 61, 114,139
reference flat, 22, 44, 63, 107, 150, 167, 183, 186
refractive index, 78, 89, 112, 152, 166, 169, 170, 175, 186
resin, 36, 42, 78, 81, 97, 121, 133, 162, 170, 177,185
resolution, 22, 43, 68, 90, 95, 126, 171, 175, 183, 186
roller coaster, 27, 32, 44
roughness, 108
rubber, 41, 94, 174

S

sapphire, 82, 160, 165, 175
saw, 8, 16, 38, 78, 87, 97, 98, 99, 100, 189
scan, 75, 81, 107, 108,
scratch, 26, 52, 57, 58, 61, 126, 188
section, 36, 67, 78, 79, 103, 188
semiconductor, 27, 29, 46, 63, 67, 70, 100, 101, 106, 108, 109, 125, 128, 145, 189
shape adjustment, 128, 129
silicon carbide, 19, 42, 50, 63, 81, 95, 103, 137, 176

silicon dioxide, 128, 168, 176
silicon nitride, 128, 131, 176
slurry, 20, 52, 62, 66, 70, 130, 140, 145
smoothing, 18, 21, 30,55, 59, 66, 96, 111, 121, 142, 177
soft metal, 29
software, 108, 135, 167
spherometer, 22
stainless steel, 29, 42, 52, 100, 146, 184
static electricity, 86
stylus, 16, 104, 105, 106, 107
sub-surface damage, 19, 59
surface finish, 16, 22, 61, 103, 105, 107, 108
surface plate, 23, 34
surface tension, 36, 39, 41, 45, 46
suspension fluid, 20, 86
sweep, 27, 31, 50, 61, 67, 105, 129, 130, 132, 139

T

temperature, 38, 39, 138, 145
thermoplasatic, 117, 127, 163, 174, 177
thermoseting, 18, 174, 177
thin layers, 8, 125
toroidal, 130
track, 23, 31, 50, 51, 130, 132
tracking, 51, 67, 177

U
undercut, 31, 35, 78, 85, 178

V
vacuum, 34, 36, 37, 38, 40, 44, 79, 185
vibration, 95

W
wafer, 37, 39, 40, 41, 45, 46, 70, 100
wafer distortion, 45
water soluble, 50, 86, 87, 189
wavelength, 16, 65, 103, 107, 188
weight, 28, 33, 34, 59, 85, 94, 131

Recent Titles in the Artech House Applied Photonics Series

Brian Culshaw and Alan Rogers, Series Editors

Chemical and Biochemical Sensing with Optical Fibers and Waveguides, Gilbert Boisdé and Alan Harmer

Coherent and Nonlinear Lightwave Communications, Milorad Cvijetic

Coherent Lightwave Communication Systems, Shiro Ryu

DWDM Fundamentals, Components, and Applications, Jean-Pierre Laude

Fiber Bragg Gratings: Fundamentals and Applications in Telecommunications and Sensing, Andrea Othonos and Kyriacos Kalli

Frequency Stabilization of Semiconductor Laser Diodes, Tetsuhiko Ikegami, Shoichi Sudo, and Yoshihisa Sakai

Handbook of Distributed Feedback Laser Diodes, Geert Morthier and Patrick Vankwikelberge

Helmet-Mounted Displays and Sights, Mordekhai Velger

Introduction to Infrared and Electro-Optical Systems, Ronald G. Driggers, Paul Cox, and Timothy Edwards

Introduction to Lightwave Communication Systems, Rajappa Papannareddy

Introduction to Semiconductor Integrated Optics, Hans P. Zappe

Liquid Crystal Devices: Physics and Applications, Vladimir G. Chigrinov

New Photonics Technologies for the Information Age: The Dream of Ubiquitous Services, Shoichi Sudo and Katsunari Okamoto editors

Optical Document Security, Third Edition, Rudolf L. van Renesse

Optical FDM Network Technologies, Kiyoshi Nosu

Optical Fiber Amplifiers: Materials, Devices, and Applications, Shoichi Sudo, editor

Optical Fiber Communication Systems, Leonid Kazovsky, Sergio Benedetto, and Alan Willner

Optical Fiber Sensors, Volume Three: Components and Subsystems, John Dakin and Brian Culshaw, editors

Optical Fiber Sensors, Volume Four: Applications, Analysis, and Future Trends, John Dakin and Brian Culshaw, editors

Optical Measurement Techniques and Applications, Pramod Rastogi

Optical Transmission Systems Engineering, Milorad Cvijetic

Optoelectronic Techniques for Microwave and Millimeter-Wave Engineering, William M. Robertson

Reliability and Degradation of III-V Optical Devices, Osamu Ueda

Signal Processing and Performance Analysis for Imaging Systems, S. Susan Young, Ronald G. Driggers, and Eddie L. Jacobs

Smart Structures and Materials, Brian Culshaw

Substrate Surface Preparation Handbook, Max Robertson

Surveillance and Reconnaissance Imaging Systems: Modeling and Performance Prediction, Jon C. Leachtenauer and Ronald G. Driggers

Wavelength Division Multiple Access Optical Networks, Andrea Borella, Giovanni Cancellieri, and Franco Chiaraluce

For further information on these and other Artech House titles, including previously considered out-of-print books now available through our In-Print-Forever® (IPF®) program, contact:

Artech House
685 Canton Street
Norwood, MA 02062
Phone: 781-769-9750
Fax: 781-769-6334
e-mail: artech@artechhouse.com

Artech House
16 Sussex Street
London SW1V 4RW UK
Phone: +44 (0)20 7596-8750
Fax: +44 (0)20 7630-0166
e-mail: artech-uk@artechhouse.com

Find us on the World Wide Web at:
www.artechhouse.com